T0174118

A HANDBOOK OF STATISTICAL GRAPHICS USING SAS ODS

A HANDBOOK OF STATISTICAL GRAPHICS USING SAS ODS

GEOFF DER

UNIVERSITY OF GLASGOW
UK

BRIAN S. EVERITT

PROFESSOR EMERITUS
KING'S COLLEGE
LONDON, UK

CRC Press
Taylor & Francis Group
Boca Raton London New York

CRC Press is an imprint of the
Taylor & Francis Group, an **informa** business

A CHAPMAN & HALL BOOK

CRC Press
Taylor & Francis Group
6000 Broken Sound Parkway NW, Suite 300
Boca Raton, FL 33487-2742

First issued in paperback 2019

© 2015 by Taylor & Francis Group, LLC
CRC Press is an imprint of Taylor & Francis Group, an Informa business

No claim to original U.S. Government works

ISBN-13: 978-1-4665-9903-1 (hbk)
ISBN-13: 978-0-367-37842-4 (pbk)

This book contains information obtained from authentic and highly regarded sources. Reasonable efforts have been made to publish reliable data and information, but the author and publisher cannot assume responsibility for the validity of all materials or the consequences of their use. The authors and publishers have attempted to trace the copyright holders of all material reproduced in this publication and apologize to copyright holders if permission to publish in this form has not been obtained. If any copyright material has not been acknowledged please write and let us know so we may rectify in any future reprint.

Except as permitted under U.S. Copyright Law, no part of this book may be reprinted, reproduced, transmitted, or utilized in any form by any electronic, mechanical, or other means, now known or hereafter invented, including photocopying, microfilming, and recording, or in any information storage or retrieval system, without written permission from the publishers.

For permission to photocopy or use material electronically from this work, please access www.copyright.com (http://www.copyright.com/) or contact the Copyright Clearance Center, Inc. (CCC), 222 Rosewood Drive, Danvers, MA 01923, 978-750-8400. CCC is a not-for-profit organization that provides licenses and registration for a variety of users. For organizations that have been granted a photocopy license by the CCC, a separate system of payment has been arranged.

Trademark Notice: Product or corporate names may be trademarks or registered trademarks, and are used only for identification and explanation without intent to infringe.

**Visit the Taylor & Francis Web site at
http://www.taylorandfrancis.com**

**and the CRC Press Web site at
http://www.crcpress.com**

Contents

Preface

Graphs, diagrams, plots etc. are essential components of almost all statistical analyses. They are needed in all stages of dealing with data, from an initial assessment of the data to suggesting what statistical models might be appropriate and for diagnosing the chosen models once they have been fitted to the data. And graphical material is often of great help to statisticians when discussing their results with clients such as psychologists, clinicians, psychiatrists and others as an aid to getting over the message the data they have collected has to tell. In this book we cover what might be termed the 'bread-and-butter' graphical methods needed in every statistician's toolkit and how to implement them using SAS Version 9.4.

SAS has two systems for producing graphs: the traditional SAS/GRAPH procedures and the newer ODS graphics. This book uses ODS graphics throughout as we believe this system offers a number of advantages. These include: their ease of use, the high quality of results, the consistency of appearance with tabular output and the convenience of semiautomatic graphs from the statistical procedures. For most users these and the new ODS graphical procedures will be all they need.

The SAS programs and the data used in this book are all available online at http://go.sas.com/hosgus/.

We hope the book will be useful for applied statisticians and others who use SAS in their work.

Chapter 1

An Introduction to Graphics: Good Graphics, Bad Graphics, Catastrophic Graphics and Statistical Graphics

1.1 The *Challenger* Disaster

January 28, 1986, was an unusually cold morning at the Kennedy Space Center (KSC) in Florida but after several days' delay the Space Shuttle *Challenger* was finally launched at 11:36 EST. There was more than the usual interest in this particular shuttle flight because of the presence on board of Christa McAuliffe, a schoolteacher who had been chosen to fly by the Teacher in Space project. A few miles from the KSC, the large crowd watching the launch that day included McAuliffe's parents and the President of the United States Ronald Reagan. Seventy-three seconds into the flight the *Challenger* broke apart leading to the deaths of its seven crew members.

The cause of the disaster was eventually traced to the O-rings that sealed the joints of the rocket motor. One of the O-ring seals had failed at liftoff allowing pressurized hot gas to escape from within the solid rocket motor leading to the eventual structural failure of the rocket; aerodynamic forces did the rest. The O-ring failure was ascribed to the low temperature at launch.

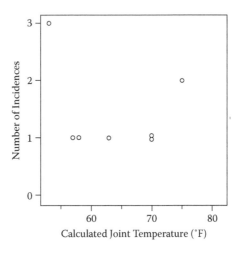

Figure 1.1   **Data plotted by space shuttle engineers the evening before the *Challenger* accident to determine the dependence of O-ring failure on temperature.**

Engineers had studied the possibility that low temperature might affect the performance of the O-rings the evening before the launch of *Challenger* by plotting data obtained from previous shuttle flights in which the O-rings had experienced thermal distress. The resulting graph (a simple *scatterplot*; see Chapter 6) is shown in Figure 1.1. The horizontal axis shows the O-ring temperature and the vertical axis shows the number of O-rings that had experienced thermal distress. The conclusion drawn by the engineers who examined this graph was that there was no relationship between temperature and incidences of thermal distress and so *Challenger* was allowed to take off when the temperature was 31°F with tragic consequences.

The data for no incidences of thermal distress were not included in the plot used by the shuttle engineers, as those involved believed that these data were irrelevant to the issue of dependence of thermal distress on temperature. They were mistaken, as shown by the plot in Figure 1.2, which includes *all* the data. Here a pattern does emerge and a dependence on temperature is revealed. Here choosing the wrong graph led to a disaster.

1.2 Graphical Displays

Just what is a graphical display? Edward Tufte gives a concise description in his now classic book, *The Visual Display of Quantitative Information*, first published in 1983:

> Data graphics visually display measured quantities by means of the combined use of points, lines, a coordinate system, numbers, symbols, words, shading and color.

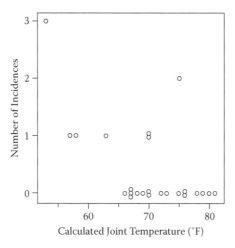

Figure 1.2   A plot of the complete O-ring data.

Graphical displays are very popular; it has been estimated (not sure by whom or how!) that between 900 billion (9×10^{11}) and 2 trillion (2×10^{12}) images of graphics are printed each year. Perhaps one of the main reasons for such popularity is that graphical presentation of data often provides the vehicle for discovering the unexpected (see Cleveland, 1993, for an example) because the human visual system is very powerful for detecting 'patterns', although it has to be remembered that whilst humans are undoubtedly good at discerning subtle patterns that are really there they are equally expert at imagining patters that are altogether absent.

Some of the advantages of graphical methods have been listed by Schmid (1954):

■ In comparison with other types of presentations, well-designed charts are more effective in creating interest and in appealing to the attention of the reader.
■ Visual relationships as portrayed by charts and graphs are more easily grasped and more easily remembered.
■ The use of charts and graphs saves time, since the essential meaning of large measures of statistical data can be visualized at a glance.
■ Charts and graphs provide a comprehensive picture of a problem that makes for a more complete and better balanced understanding than could be derived from tabular or textual forms of presentation.
■ Charts and graphs can bring out hidden facts and relationships and can stimulate, as well as aid, analytical thinking and investigation.

Schmid's last point is reiterated by the legendary John Tukey in his observation that 'the greatest value of a picture is when it forces us to notice what we never expected to see'.

The prime objective of a graphical display is to communicate to others and ourselves. Graphic design must do everything it can to help people understand. In some cases a graphic is required to give an overview of the data and perhaps to tell a story about the data. In other cases a researcher may want a graphical display to suggest possible hypotheses for testing on new data and after some model has been fitted to the data a graphic that criticizes the model may be what is needed. Examples of using graphic displays in all these situations will be found throughout this text. But for now we will consider a little history and some early examples of graphical displays.

1.3 A Little History and Some Early Graphical Displays

Conveying complex information in words has always been considered difficult, and alternative methods of communication have been sought. Wainer (1997) points out that as far back as preclassical antiquity, paleolithic art provided an early and very striking example of graphic display with carvings of animals being intermixed with patterns of dots and strokes that archeologists have interpreted as a lunar notation system related to the animal's seasonal appearance. And Egyptian geographers turned spatial information into spatial diagrams and maps to keep track of land shifted by river floods.

But we have to make a leap into the 17th and 18th centuries to meet some early graphics that are vaguely familiar to us. William Playfair, for example, is often credited with inventing the *bar chart* (see Chapter 3) in the last part of the 18th century, although a Frenchman, Nicole Oresme, used a bar chart in a 14th century publication, *The Latitude of Forms*, to plot velocity of a constantly accelerating object against time. But it was Playfair who popularized the idea of graphic depiction of quantitative information. Figure 1.3 shows one of Playfair's earliest bar charts used to show imports and exports of Scotland.

Playfair's graphs are essentially one-dimensional and the next major graphical invention and the first truly two-dimensional graph is the *scatterplot* (or *scattergram*) an example of which introduced this chapter. According to Friendly and Denis (2006) the humble scatterplot may be considered the most versatile, polymorphic and generally useful invention in the entire history of statistical graphics. As we have already seen in Section 1.1, a scatterplot is quintessentially a plot of two variables x and y measured independently to produce bivariate pairs (x_i, y_i) and displayed as individual points of a coordinate grid typically defined by horizontal and vertical axes, where there is no necessary functional relation between x and y. It is worthwhile here quoting Tufte's comment about the scatterplot (1983):

> ...the scatterplot and its variants is the greatest of all graphical designs. It links at least two variables encouraging and even imploring the viewer to assess the possible causal relationship between the plotted variables. It confronts causal theories that x causes y with empirical evidence as to the actual relationship between x and y.

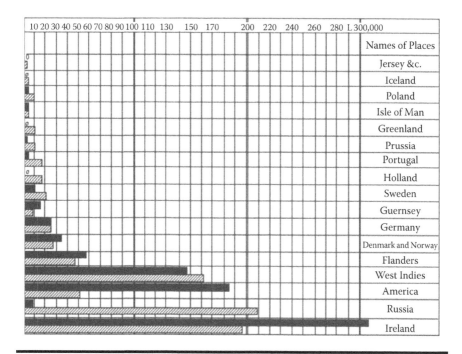

Figure 1.3   Playfair's bar chart for imports and exports of Scotland: imports are represented by crosshatch bars and exports by solid bars.

There are various claimants for the first genuine scatterplot, but Friendly and Denis (2006) come down on the side of Francis Galton's graphical displays constructed in his work on correlation, regression and heritability, although these are somewhat less than true scatterplots of data as used today being essentially bivariate frequency tables. The earliest known example of one of these charts is shown in Figure 1.4.

Friendly and Denis (2006) call diagrams such as Figure 1.4 'a poor-man's scatterplot' but note that for Galton such diagrams allowed him to 'smooth' the numbers by averaging the four adjacent cells to produce a simple *bivariate density estimator* (see Chapter 6), a procedure which eventually led to diagrams such as Figure 1.5, a graphic that led on to a host of important developments in statistics, for example, correlation, regression and partial correlation.

Another famous scatterplot which led to insights into how stars could be classified is the *Hertzsprung–Russell diagram* (H–R diagram), pioneered independently by Elnar Hertzsprung and Henry Norris Russell in the early 1900s. The H–R diagram is a plot of the luminosity of a star against its temperature; an example is given in Figure 1.6 and shows that stars preferentially fall into certain regions of the diagram with the majority falling along a curving diagonal line called the *main sequence*. The significance of the H-R diagram when

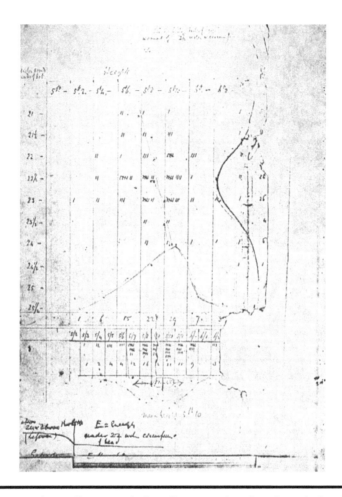

Figure 1.4 **Galton's first correlation diagram, showing the relation between head circumference and height, from his undated notebook 'Special Peculiarities'. (From Hilts, V. L., 1975, *A Guide to Francis Galton's English Men of Science*, Philadelphia, American Philosophical Society, Figure 5, p. 26. With permission.)**

it was first plotted is that stars were seen as concentrated in distinct regions rather than being distributed at random.

Lastly in this section we will consider a plot made in the 19th century by an early epidemiologist, John Snow (1813–1858), which was probably responsible for saving many lives. After an outbreak of cholera in central London in September 1854, Snow used data collected by the General Register Office and plotted the location of deaths on a map of the area and also showed the location of the area's eleven water pumps. The resulting map is shown in Figure 1.7 (deaths are marked

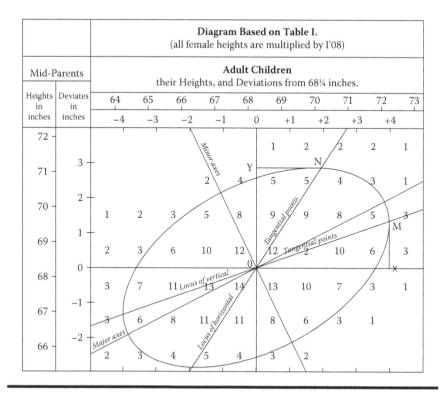

Mid-Parents		Diagram Based on Table I. (all female heights are multiplied by 1'08)									
		Adult Children their Heights, and Deviations from 68¼ inches.									
Heights in inches	Deviates in inches	64	65	66	67	68	69	70	71	72	73
		−4	−3	−2	−1	0	+1	+2	+3	+4	

Figure based on the diagram showing:

- 72 —
- 71 — 3
- 70 — 2
- 69 — 1
- 68 — 0
- 67 — −1
- 66 — −2

Values within ellipse including labels *Minor axes*, Y, N, M, *Tangential points*, *Locus of vertical*, *Locus of horizontal*, *Major axes*, and the numbers:

1 2 2 2 1
2 4 5 5 4 3 1
1 2 3 5 8 9 9 8 5 3
2 3 6 10 12 12 2 10 6 3
0
3 7 11 13 14 13 10 7 3 1
3 6 8 11 11 8 6 3 1
2 3 4 5 4 3 2

Figure 1.5 Galton's smoothed correlation diagram for the data on heights of parents and children, showing one ellipse of equal frequency. (From Galton, F., 1886, Regression towards mediocrity in hereditary stature, *Journal of the Anthropological Institute of Great Britain and Ireland*, 15, 246–263, Plate X. With permission.)

by dots and water pumps by crosses). Examining the scatter over the surface of the map, Snow observed that nearly all the cholera deaths were among those who lived near the Broad Street pump. But before claiming that he had discovered a possible causal connection, Snow made a more detailed investigation of the deaths that had occurred near some other pump. He visited the families of ten of the deceased and found that five of these, because they preferred its taste, regularly sent for water from the Broad Street pump. Three others were children who attended a school near the Broad Street pump. One other finding that initially confused Snow was that there were no deaths amongst workers in a brewery close to the Broad Street pump, a confusion that was quickly resolved when it became apparent that the workers drank only beer, never water! Snow's findings were sufficiently compelling to persuade the authorities to remove the handle of the Broad Street pump and in days the neighbourhood epidemic that had taken more than 500 lives had ended.

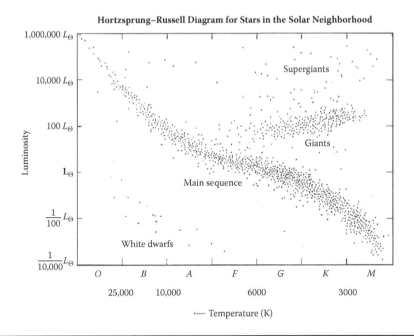

Figure 1.6   Herztsprung–Russell diagram.

1.4 Graphical Deception

Not all graphical displays are as honest as they should be and it is relatively easy to mislead the unwary with graphical material. For example, consider the plot of the death rate per million from cancer of the breast for several periods over the last three decades, shown in Figure 1.8. The rate appears to show a rather alarming increase. However, when the data are replotted with the vertical scale beginning at zero, as shown in Figure 1.9, the increase in the breast cancer death rate is altogether less startling. This example illustrates that undue exaggeration or compression of the scales is best avoided when drawing graphs (unless, of course, you are actually in the business of deceiving your audience).

A very common distortion introduced into the graphics most popular with newspapers, television and the media in general is when *both* dimensions of a *two-dimensional figure* or *icon* are varied simultaneously in response to changes in a single variable. The examples shown in Figure 1.10, both taken from Tufte (1983), illustrate this point. Tufte quantifies the distortion with what he calls the *lie factor* of a graphical display, which is defined as the size of the effect shown in the graph divided by the size of the effect in the data. Lie factor values close to unity show that the graphic is probably representing the underlying numbers reasonably accurately. The lie factor for the oil barrels is 9.4 since a 454% increase is depicted as 4280%. The lie factor for the shrinking doctors is 2.8.

Figure 1.7 Snow's map of cholera deaths in the Broad Street area.

A further example given by Cleveland (1994) and reproduced here in Figure 1.11 demonstrates that even the manner in which a simple scatterplot is drawn can lead to misperceptions about data. The example concerns the way in which judgment about the correlation of two variables made on the basis of looking at their scatterplot can be distorted by enlarging the area in which the points are plotted. The coefficient of correlation in the lower diagram in Figure 1.11 appears greater.

Some suggestions for avoiding graphical distortion taken from Tufte (1983) are

■ The representation of numbers, as physically measured on the surface of the graphic itself, should be directly proportional to the numerical quantities represented.

■ Clear, detailed and thorough labelling should be used to defeat graphical distortion and ambiguity. Write out explanations of the data on the graphic itself. Label important events in the data.

■ To be truthful and revealing, data graphics must bear on the heart of quantitative thinking: Compared to what? Graphics must not quote data out of context.

■ Above all else show the data.

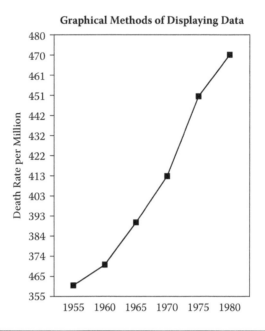

Figure 1.8 Death rates from cancer of the breast where the *y*-axis does not include the origin.

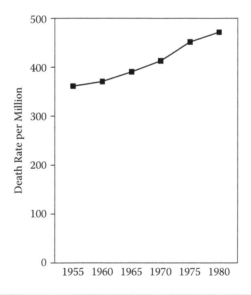

Figure 1.9 Death rates from cancer of the breast where the *y*-axis does include the origin.

(a) (b)

Figure 1.10   Graphics exhibiting lie factors of (a) 9.4 and (b) 2.8.

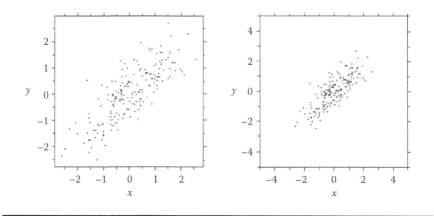

Figure 1.11   Misjudgment of size of correlation caused by enlarging the plot area.

In the earlier examples the implication was that the producer of the graphs was deliberately trying to pull the wool over the eyes of the viewer, but there are examples of graphs in the literature which are simply poor at communicating the message contained in the data from which they are produced and which, if replotted in some other way, would give a far clearer picture of the message in the data.

To illustrate this type of situation we can look at an example of diagram given in Vetter (1980) and shown here in Figure 1.12. The aim of the diagram is to display the percentages of degrees awarded to women in several disciplines of science and technology during three time periods. At first glance the labels on the diagram suggest that the graph is a standard divided bar chart with the length of the bottom division of each bar showing the percentage for doctorates, the length of the middle division showing the percentage for master's degrees and the top division showing the percentage for bachelor's degrees. A little reflection shows that this interpretation is not correct, since it would imply that in most cases the percentage of bachelor's degrees given to women is lower than the percentage of doctorates. Closer examination of the diagram reveals that the three values of the data for each discipline during each time period are determined by the three adjacent vertical dotted lines. The top of the left-hand line indicates the value for doctorates, the top end of the middle line indicates the value for master's degrees and the top end of the right-hand line indicates the value for bachelor's degrees.

Cleveland (1994) discusses the diagram in Figure 1.12 and points out that the manner of the diagram's construction makes it hard to connect visually the three values of a particular type of degree for a specific discipline, thus making it difficult to see changes over time. Figure 1.13 shows the data represented by Figure 1.12

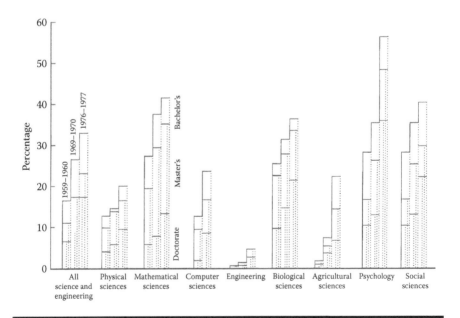

Figure 1.12 **Proportion of degrees in science and engineering earned by women in the periods 1959–1960, 1969–1970 and 1976–1977. (Reproduced with permission from Vetter, B. M., 1980, Working women scientists and engineers, *Science*, 207, 28–34. ©1980 American Association for the Advancement of Society.)**

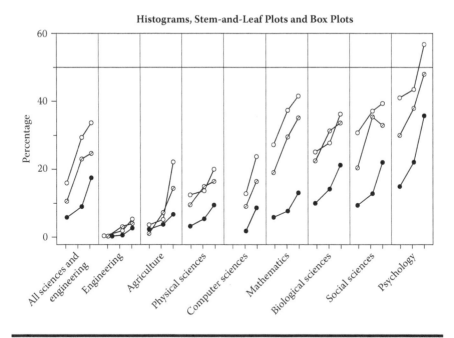

Figure 1.13 Percentage of degrees earned by women for three degrees (⊙ bachelor's degree; ⊗ master's degree; • doctorate), three time periods and nine disciplines. The three points for each discipline and degree indicate the periods 1959–1960, 1969–1970, 1976–1977.

replotted by Cleveland in a bid to achieve greater clarity. It is now clear how the data are represented, and this diagram allows viewers to easily see the percentages corresponding to each degree, in each discipline, over time. Finally the figure caption explains the content of the diagram in a comprehensive and clear fashion. All in all Cleveland appears to have produced a graphic that would satisfy even that doyen of graphical presentation, Edward R. Tufte, in his demand that 'excellence in statistical graphics consists of complex ideas communicated with clarity, precision and efficiency'.

Wainer (1997) gives a further demonstration of how displaying data as a bar chart can disguise rather than reveal important points about data. Figure 1.14 shows a bar chart of life expectancies in the middle 1970s divided by sex for ten industrialized nations. The order of presentation is alphabetical (with the USSR positioned as Russia). The message we get from this diagram is that there is little variation and women live longer than men. But displaying the data in the form of a simple *stem-and-leaf plot* (see Figure 1.15) the magnitude of the sex difference (seven years) is immediately clear as is the unusually short life expectancy for men in the USSR, whereas Russian women have similar life expectancy to women in other countries.

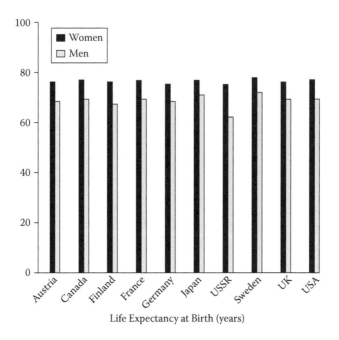

Life Expectancy at Birth (years)

Figure 1.14   Life expectancy at birth by sex and by country.

Women	Age	Men
Sweden	78	
France, UD, Japan, Canada	77	
Finland, Austria, UK	76	
USSR, Germany	75	
	74	
	73	
	72	Sweden
	71	Japan
	70	
	69	Canada, UK, US, France
	68	Germany, Austria
	67	Finland
	66	
	65	
	64	
	63	
	62	USSR

Figure 1.15   An alternative display of life expectancies at birth by sex and by country.

1.5 Summary

The use of graphical displays in general and in statistics in particular is widespread. The increase in the use of such display has clearly paralleled advances in computing, because computers can supply a vast variety of diagrams and plots rapidly and accurately. So because the machine is doing the work the question is no longer shall we plot but rather what shall we plot. The options are almost endless and include interactive and dynamic graphics (see, for example, Cleveland and McGill, 1987, and Cook and Swayne, 2007) and new approaches to graphics for large datasets are described in Unwin, Theus and Hofmann (2006). In this book we will show how to produce many useful statistical graphs; some for giving an overview of a dataset and perhaps to tell a story about the data, others that might suggest possible hypotheses that could be tested on new data and yet others that can be used to diagnose and criticize fitted statistical models. All the graphs and diagrams that appear in the book have been produced by the SAS ODS graphics and in the next chapter we give an overview of this feature of SAS.

Chapter 2

An Introduction to ODS Graphics

2.1 Generating ODS Graphs

There are four methods of producing ODS (output delivery system) graphics: (i) the ODS Graphics Designer, an interactive application within SAS that uses a point and click interface to create graphs; (ii) the statistical graphics procedures, sgplot, sgpanel and sgscatter; (iii) ODS graphs produced semi-automatically by statistical procedures; and (iv) programs written in the graphical template language. We concentrate on the second and third of these in the belief that these will furnish the majority of users with all they need. Users who become familiar with this material will also find the ODS Graphics Designer easy to use, but the reverse will not necessarily be true. The fourth option, directly programming in the graphical template language, offers the greatest flexibility since that underlies the whole of ODS graphics but is a more complex approach. For those wishing to go on to that the material, here will make a good starting point.

2.1.1 SAS Settings for ODS Graphics

In order to produce ODS graphics two conditions must be met: an ODS destination must be open and ODS graphics must be enabled. The first of these two is generally unproblematic since at least one ODS destination needs to be open in order to view the output from SAS procedures. If no ODS destination is open a warning will appear in the SAS log.

The status of ODS graphics and the ODS destination in use when SAS starts can be viewed and changed from the Tools menu by selecting Options, Preferences,

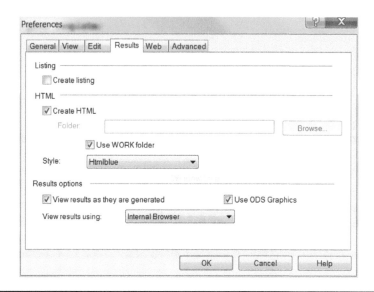

Figure 2.1   Window showing default SAS system options for results.

then clicking on the Results tab. Figure 2.1 shows the default settings under SAS version 9.4. Use ODS Graphics is selected, meaning that ODS graphics are enabled. Create HTML is selected and Create listing is not. Note also that View results as they are generated is also selected.

HTML became the default destination in version 9.3; prior to that listing was the default. At the same time ODS graphics were enabled by default. Although Figure 2.1 shows the default, not all installations of SAS will have been set up in this way and it is worth consulting the menu to check. We also believe that it is easier while learning to use SAS and the ODS graphics to revert to the older defaults, so recommend unsetting Create HTML and setting Create listing and unsetting View results as they are generated.

Whilst ODS graphics may be enabled and disabled via the Use ODS Graphics option in this window, it is generally more convenient to use the ods graphics on; and ods graphics off; statements as this gives control over other aspects of ODS graphics. We will return to the ods graphics statement later.

2.2 ODS Destinations

Before turning to ODS graphics it is worth considering some relevant aspects of ODS and ODS destinations.

ODS is an abbreviation for Output Delivery System and is the part of SAS that controls how the output from SAS procedures is formatted. For each type of formatting to be applied to SAS output there is a corresponding ODS 'destination'. The most commonly used ODS destinations include:

html (hypertext markup language) – The formatting system widely used for web pages and the current default in SAS.

listing – The plain text format which is displayed in the SAS output window, the default prior to version 9.3.

pdf (portable document format) – The format originally devised by Adobe and now widely used for documents that are designed to be readable on a wide range of computer systems and devices.

rtf (rich text format) – A format designed to be compatible with word processing packages, especially Microsoft Word.

powerpoint – The format of 'slides' for inclusion in Microsoft PowerPoint presentations.

ODS destinations are opened and closed, respectively, with statements of the form:

```
ods <destination>;
```

and

```
ods <destination> close;
```

While an ODS destination remains open, all the output from procedures is routed to that destination. This includes both tabular and graphical output but ODS destinations differ in the way that tables and graphs are stored. For some destinations, such as rtf and pdf, the graphs are included in the same document as the tables, whereas for others, including html and listing, the graphs are stored separately from the tables and with a separate file for each graph.

While learning about ODS graphics it is more convenient to have the graphical and tabular output separate. Both the listing and html destinations keep the underlying files separate, but they are more easily handled separately using the listing destination; hence our preference for using it.

The default location for the graph files is the current active directory (the one listed at the bottom right of the SAS screen) but a different directory can be specified for graphs by using the gpath option of the ODS statement for the destination, for example:

```
ods html gpath='c:\mygraphs';
ods listing gpath='c:\mygraphs';
```

2.3 Statistical Graphics Procedures

There are five statistical graphics procedures that may be used to produce ODS graphics: sgplot, sgpanel, sgscatter, sgdesign and sgrender.

2.3.1 Proc Sgplot

Proc sgplot is a very flexible plotting procedure that can produce a wide range of different plot types and is useful for both routine and bespoke graphics. It is the workhorse of ODS graphics and, apart from the graphs available in the statistical procedures, many users will never need to use anything else. For this reason it is well worth investing time in becoming familiar with its full range of capabilities.

To illustrate the basics of proc sgplot we will use a small dataset giving some measurements of percentage body fat for 18 subjects together with their age and sex. The data are shown in Table 2.1.

Table 2.1 Data on Body Composition by Age and Sex

Age	% Body Fat	Sex
23	9.5	M
23	27.9	F
27	7.8	M
27	17.8	M
39	31.4	F
41	25.9	F
45	27.4	M
49	25.2	F
50	31.1	F
53	34.7	F
53	42.0	F
54	29.1	F
56	32.5	F
57	30.3	F
58	33.0	F
58	33.8	F
60	41.1	F
61	34.5	F

2.3.1.1 xy Plots

An *xy* plot is one in which the data are represented in two dimensions defined by the values of two variables. The use of plots of this type will be illustrated and discussed in detail in later chapters. Here we want to show how to produce them with **proc sgplot** and some of the variations that are possible.

The simplest *xy* plot is a scatterplot and can be illustrated using the body composition dataset.

```
proc sgplot data=bodycomp;
  scatter y=pctfat x=age;
run;
```

The syntax is straightforward: a **scatter** statement is used and the **x** and **y** variables specified explicitly. As usual, the dataset used is named on the **proc** statement. For different types of plots a statement other than **scatter** is used. Table 2.2 shows some *xy* plots that could be generated by **sgplot**. Most of these will be illustrated in later chapters.

Table 2.2   *xy* Plots Using Sgplot

Plotting Statement	Type of Plot
scatter	Scatterplot – Data values plotted
series	Line plot – Data values joined with lines
step	Step plot – Data values joined with stepped lines
needle	Needle plot – A scatterplot with a line joining the value to the baseline
reg	Regression plot – A scatterplot with a regression line, or curve fitted
loess	As reg but with a locally weighted regression curve added
pbspline	As reg with penalised Beta spline smoothing curve fitted
bubble	Bubble plot – A scatterplot with bubbles that are proportional in size to the value of a third variable
highlow	High low close – A set of lines each joining two values on one of the axes with an optional mark at a third value
band	Band – A band, usually a confidence band
vector	Vector – For plotting arrows or lines at x,y locations

For line plots and step plots the points will be plotted in the order in which they occur in the dataset, so it is usually necessary to sort the data by the *x*-axis variable first.

Each of these plot types has a set of options, which may be specified after a slash (/). Many of the options are specific to the type of plot, but common and useful options are

group=<var>	Uses different markers/lines/fills for each value of <var>
datalabel=<var>	Uses the values of <var> to label data points
markerchar=<var>	Use the values of <var> as markers
jitter	Offsets data points that have the same values
transparency	Specifies how transparent the plot will be
nomarkers	Suppresses markers
markers	Adds markers

2.3.1.2 Grouped Plots

Of the options listed above the **group=** option is the most important and widely used. It distinguishes different subgroups of the dataset within the plot and is typically used to explore or illustrate differences between the groups. The subgroups are defined by the values of the variable specified on the option. This type of plot is often referred to as a 'grouped' plot for obvious reasons. Its use can be simply illustrated as follows:

```
proc sgplot data=bodycomp;
  reg y=pctfat x=age/group=sex;
run;
```

The result is shown in Figure 2.2. In the plot males and females are distinguished by different markers, circles for males and pluses for females; separate regression lines have been fitted for each sex and these are plotted with different line types.

When the **group=** option has been used a legend is produced as a key to how the groups are plotted. If preferred, this legend can be suppressed with the **noauto-legend** option on the **proc sgplot** statement.

2.3.1.3 Attribute Options

There is a further set of options designed to control the appearance of parts of a graph such as its lines, markers, curves, shaded areas etc. These options have

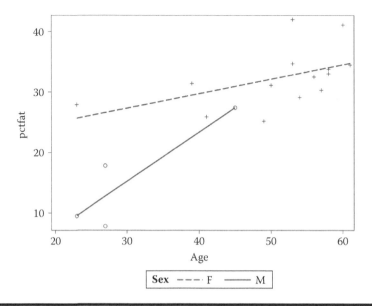

Figure 2.2   Grouped regression plot.

names such as lineattrs, markerattrs and fillattrs, where attrs is an abbreviation of attributes and the first part of the name corresponds to the plot element whose appearance is controlled. For lines, the colour, thickness and pattern can be specified, so for a red, dashed line, 10 pixels thick the option would be

```
lineattrs=(color=red pattern=dash thickness=10)
```

The attributes and their values are within parentheses. It is not necessary to specify all three attributes.

Available line patterns include: Solid, Dash, Dot, ShortDash, MediumDash, LongDash, plus various combinations of dashes and dots, such as DashDashDot, DashDotDot, ShortDashDot LongDashShortDash etc.

The default units for thickness, as for size in general, are pixels (px) so the example is equivalent to thickness=10px. Other size units are centimetres (cm), inches (in), millimetres (mm) percentage (pct or %) and point (pt).

2.3.1.4 Overlaid Plots

It is often useful to combine the information from two or more plots by overlaying them. Proc sgplot does this automatically when more than one plotting statement is included. For example, a plot to compare the fits from linear and locally weighted regression could be produced as follows (locally weighted regression is explained in Chapter 6):

```
proc sgplot data=bodycomp;
   reg y=pctfat x=age;
   loess y=pctfat x=age/nomarkers;
run;
```

The nomarkers option is specified on the loess statement to prevent the data points being plotted twice (i.e. once each by the reg and loess statements) as sgplot uses different plotting symbols for each.

The basic *xy* plot can be enhanced with confidence bands (band) or lines (highlow or vector) and reference lines (refline). When used in conjunction with some data processing very sophisticated plots can be produced.

2.3.1.5 Summary Plots

Plots of summary statistics are often useful when comparing groups. Proc sgplot can produce plots of various summary statistics as a bar plot, line plot or dot plot. The plot statements are vbar/hbar and vline/hline, depending on whether vertical or horizontal orientation is desired, and dot. To illustrate this, the variable age in the bodycomp dataset is first recoded into ten-year bands.

```
data bodycomp;
   set bodycomp;
   decade=int(age/10)*10;
run;
```

```
proc sgplot data=bodycomp;
   vline decade/response=pctfat stat=mean limitstat=stddev;
run;
```

The vline statement requires only one variable, decade in this case, which provides the values for the *x* axis, referred to in this case as the *category* axis as categorical variables can be used. The rest of the vline statement in this example consists of options. Without these options the statement will result in a plot of the frequency count within each decade. The options used here specify a response variable, pctfat, whose summary values will provide the *y* axis. The stat= option specifies the mean as the summary statistic; other possibilities are freq, median, percent and sum. Limits of one standard deviation are also to be drawn on the plot (limitstat=stddev); the other possibilities are clm for confidence limits and stderr for standard errors. When the limitstat value is stddev or stderr, the numstd= option can be used to specify the number of standard deviations or standard errors to be plotted. The default is 1. The result is shown in Figure 2.3. There is no limit line for decade 30 as there is only one person in this age range.

The **vbar** and **hbar** statements work in a very similar way to **vline** and **hline** and can produce a wide range of bar plots, particularly when used with the **group** option as in the following example.

```
proc format;
   value age2grp low-50='younger' 50<-high='older';
   value $sex 'F'='Female' 'M'='Male';
run;

proc sgplot data=bodycomp;
   vbar sex / response=pctfat stat=mean limitstat=stderr
              limits=upper group=age groupdisplay=cluster;
   format age age2grp. sex $sex.;
run;
```

We first create two formats, one for a median split of age and the other to give better labelling of **sex**. On the **vbar** statement **sex** defines the category axis; the **response** and **stat** options are as in the previous example but this time the standard error is used for the limits and only the upper limit is to be displayed. The formatted value of **age** is used for subgroups and these are to be displayed as a clustered bar chart. The default value for **groupdisplay** is **stack**, which would give a stacked bar chart. The resulting plot is shown in Figure 2.4.

Dot plots are illustrated in the next chapter.

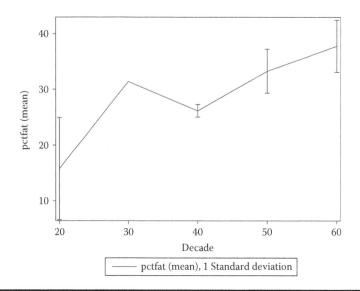

Figure 2.3   Plot of means and standard deviations using vline statement.

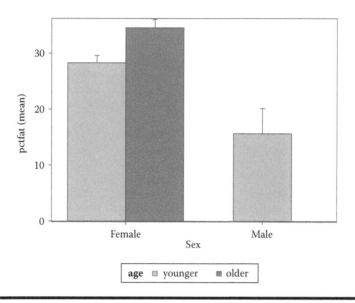

Figure 2.4　Clustered bar plot.

2.3.1.6 Distribution Plots

The histogram, density, hbox and vbox statements plot the distribution of a continuous variable as a histogram, density plot or box plot (horizontal or vertical). These plots are described in detail in the next chapter.

For categorical variables the vbar and hbar statements may be used without a response variable either with stat=freq, which is the default and so can be omitted, or with stat=percent. For example:

```
proc sgplot data=bodycomp;
   vbar decade;
run;
```

```
proc sgplot data=bodycomp;
   vbar decade/stat=pct;
run;
```

These sgplot steps produce identical results but with different labelling of the *y* axis. The result of the second is shown in Figure 2.5.

Another commonly used bar chart shows the percentage breakdown within groups. Using the two formats defined earlier we can illustrate this type of plot as follows:

```
proc sgplot data=bodycomp pctlevel=group;
   vbar sex/stat=pct group=age barwidth=.5;
   format age age2grp. sex $sex.;
run;
```

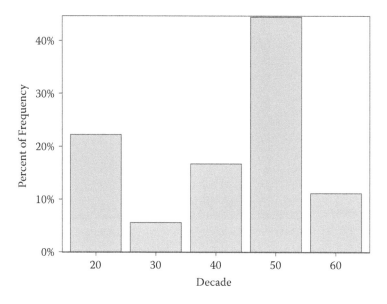

Figure 2.5   Bar chart showing percentage breakdown of age.

The pctlevel=group option on the proc statement specifies that the percentages are to be calculated across groups within a category so that the category sums to 100. As mentioned earlier the default for vbar/hbar is groupdisplay=stack so this option has been omitted. The barwidth option has been used to reduce the width of the bars. Values for this option range from 0 to 1 and give the proportion of the overall width occupied by the bars, the remainder being the space between the bars. For a small number of categories the default barwidth of .8 gives rather fat looking bars. The result is shown in Figure 2.6.

Proc sgplot is used extensively throughout the book and many more options are illustrated in subsequent chapters.

2.3.2 Panel Plots

A panel plot is a graph containing two or more plots within it. The reason for combining plots in a panel is either that they share a common purpose or to make comparisons between them.

Proc sgpanel and proc sgscatter both produce multiple plots contained within a grid of panels. Within sgpanel the grid is defined by the values of one or more variables in the dataset with the result that each plot contains a subset of the data. With sgscatter each plot contains the full set of data and the grid is an arrangement of pairs of variables, with or without common axes.

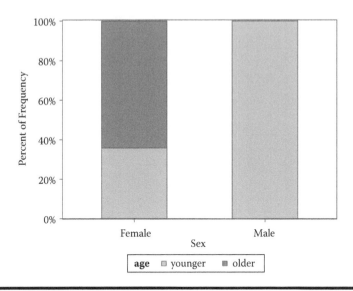

Figure 2.6 Bar chart showing percentage breakdown by age for males and females.

2.3.2.1 Proc Sgpanel

The **sgpanel** procedure functions very similarly to **sgplot**. The main difference is the **panelby** statement. This specifies the variables which are to be used to divide the data into the subsets that will be plotted separately within panels. We can illustrate this with a simple example using the **bodycomp** dataset and with **sex** used to define the panels.

```
proc sgpanel data=bodycomp;
   panelby sex;
   reg y=pctfat x=age;
run;
```

Apart from the **panelby** statement the syntax is the same as would be used in **proc sgplot**. The result is shown in Figure 2.7. It is worth comparing this with Figure 2.2 as the two contain the same information.

The **panelby** statement has a number of options (after a /) to control how the panels are arranged. The number of rows and columns in the panel is controlled by the **rows=** *n* and **columns=** *n* options. If the previous example were re-run with **panelby sex/rows=2;** the panels would be stacked one above the other. The **spacing=** option adds some space between the panels. The header for each panel takes the form **variable name=value**. The **novarname** option gives just the values as headers. If the **novarname** option had been used in the earlier examples the headers of the panels would have contained simply 'F' and 'M'. To suppress the header altogether use the **noheader** option. In Figure 2.7 the *x* and *y* axes have the same range in the two plots

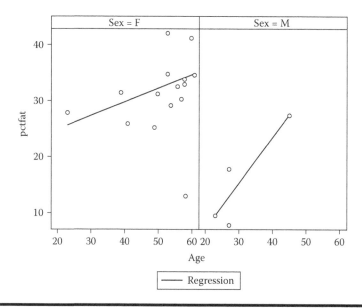

Figure 2.7   Panel plot showing regression of percent body fat on age separately for males and females.

even though the range of data values for males is much smaller than for females. This behavior is controlled by the **uniscale=** option with values **column, row** and **all** (the default).

Nearly all the plot statements available in **proc sgplot** are also available in **proc sgpanel**.

2.3.2.2 Proc Sgscatter

The **scatter** statement within **proc sgplot** produces a scatterplot, whereas **proc sgscatter** produces a grid of scatter plots commonly referred to as a scatterplot matrix. The use of these is described in later chapters. Here we introduce the syntax using a subset of Fisher's Iris data.

```
data iris2;
  set sashelp.iris;
  if species~='Setosa';
run;

proc sgscatter data=iris2;
    matrix sepallength sepalwidth petallength petalwidth
         / group=species diagonal=(histogram normal);
run;
```

The **matrix** statement names a list of variables that go to make up the scatter-plot matrix, which comprises a grid of scatterplots, one for each pair of variables.

The two options specify that the species are to be distinguished in the plot and histograms with normal curves are to be plotted on the diagonal. The default is for the diagonal to contain the variable names. The result is shown in Figure 2.8.

A second way in which the pairs of variables to be plotted can be specified is by a compare statement, which lists one or more *x* variables and one or more *y* variables. If there is more than one *x* or *y* variable, the names are enclosed in parentheses. An example would be

```
compare x=(sepallength sepalwidth) y=(petallength petalwidth)/group=species;
```

When compare is used row and column axes are shared.

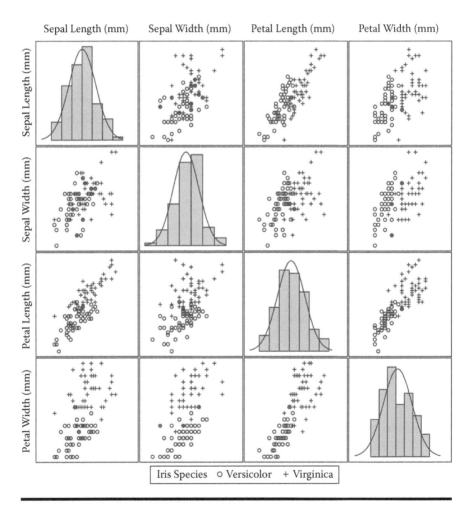

Figure 2.8 Scatterplot matrix of measurements on two species of iris.

The third way is via a **plot** statement whereby *xy* pairs are specified in the form *y*x* and again multiple variables can be enclosed in parentheses. For example:

```
plot sepallength*sepalwidth petallength*petalwidth/group=species;
```

In contrast to the **compare** statement each plot has its own set of axes and the plots can be completely independent.

2.3.3 Proc Sgdesign

Proc sgdesign produces a graph from a specification made using the ODS Graphics Designer, an interactive application within SAS that uses a point and click interface to create graphs.

2.3.4 Proc Sgrender

Proc sgrender is for bespoke plots programmed in the graphics template language which underlies the statistical graphics procedures. A detailed description is beyond the scope of this book but a simple example that could be easily adapted is given in Chapter 6.

2.4 ODS Graphs from Statistical Procedures

Most statistical procedures are capable of directly producing ODS graphs for use in checking the assumptions of the procedure or interpreting and presenting the results.

The graphs that are available for each procedure are listed in the help section for the procedure under the **Details** tab as **ODS Graphics**. This section of the help varies according to the procedure, but typically contains some general descriptions of the plots, or types of plot, that are available and the syntax or options that are required to produce them. This information is usually summarized in a table. Table 42.14 of the SAS documentation has been reproduced here as Table 2.3 and shows the summary table of ODS graphics produced by **proc glm**. The column headed Plot Description is self-explanatory, but the other two columns merit some explanation. The ODS Graph Name is the name the graph is referred to by the ODS system. If the ODS destination is one that produces separate files for tabular output and graphs, as the html destination does, the name of the file containing the graph will be based on this name with an extension corresponding to output format of the graph, .jpg for JPEG, .ps for postscript etc. Where more than one graph of the same name is produced a numerical suffix is added to the name and incremented. If three jpeg format fitplots were produced they would be named fitplot.jpg, fitplot1.jpg and fitplot2.jpg.

Table 2.3   SAS Help Table Showing ODS Graphics Available for Proc Glm

ODS Graph Name	Plot Description	Option
ANCOVAPlot	Analysis of conversion plot	Analysis of covariance model
ANOMPlot	Plot of LS-mean differences against average LS-mean	LSMEANS / PDIFF=ANOM
BoxPlot	Box plot of group means	One-way ANOVA model or MEANS statement
ContourFit	Plot of predicted response surface	Two-predictor response surface model
ControlPlot	Plot of LS-mean differences against a control level	LSMEANS / PDIFF=CONTROL
DiagnosticsPanel	Panel of summary diagnostics for the fit	PLOTS=DIAGNOSTICS
CooksDPlot	Cooks D plot	PLOTS=DIAGNOSTICS(UNPACK)
ObservedByPredicted	Observed by predicted	PLOTS=DIAGNOSTICS(UNPACK)
QQPlot	Residual Q-Q plot	PLOTS=DIAGNOSTICS(UNPACK)
ResidualByPredicted	Residual by predicted values	PLOTS=DIAGNOSTICS(UNPACK)
ResidualHistogram	Residual histogram	PLOTS=DIAGNOSTICS(UNPACK)
RFPlot	RF plot	PLOTS=DIAGNOSTICS(UNPACK)
RStudentByPredicted	Studentized residuals by predicted	PLOTS=DIAGNOSTICS(UNPACK)
RStudentByLeverage	RStudent by hat diagonals	PLOTS=DIAGNOSTICS(UNPACK)
DiffPlot	Plot of LS-mean pairwise differences	LSMEANS / PDIFF
IntPlot	Interation plot	Two-way ANOVA model
FitPlot	Plot of predicted response by predictor	Model with one continuos predictor
ResidualPlots	Plots of the residuals against each continuos covariate	PLOTS=RESIDUALS

The ODS name may also be used to refer to a plot in other ODS statements. If a procedure produces several diagnostic plots but we want to see only the Q-Q plot, we could include the statement

ods select qqplot;.

The Option column in Table 2.3 shows the conditions under which the plot is produced by the procedure. Some plots need to be requested by using a specific procedure option, very often this is the **plots=** option. Some plots are produced by default but may depend on the type of model or analysis that is being performed. So for **proc glm**: an ANCOVAplot, BoxPlot, Contourfit, IntPlot or FitPlot will be produced if the appropriate, corresponding type of model is fitted; diagnostics and residuals plots need to be invoked with the **plots=** option; whereas a ControlPlot or DiffPlot require an **lsmeans** statement within the proc step with the **pdiff** option.

2.4.1 *plots= Option*

The most common way in which ODS graphs are invoked in statistical procedures is via the **plots=** option. This is usually on the **proc** statement, but may be on other statements usually when more than one statement has a **plots=** option. In these cases, there will be an additional column in the summary table showing which statement has the relevant option.

Although the plots that are available vary from one procedure to another, the **plots=** option has a common syntax with two general forms:

```
PLOTS <(global-plot-options)> <=plot-request <(options)>>
PLOTS <(global-plot-options)> <=(plot-request <(options)> <... plot-request
     <(options)>>)>
```

These are best explained by some examples.

2.4.2 *Single Plot Request*

The simplest form is a single plot request, such as

```
plots=all
plots=none
plots=residuals
```

A single plot request may result in more than one plot, as in **plots=all**, or no plots at all as in the admittedly unusual case of **plots=none**. A single plot request may specify options for that plot and these are given in parentheses after the plot name:

```
plots=residuals(smooth)
```

Some plots come as a group in a panel plot and these have an option to unpack the panel into separate graphs. So, in the case of **proc glm**, **plots=diagnostics** will produce a panel of diagnostic plots with the ODS name **DiagnosticsPanel**, whereas **plots=diagnostics(unpack)** will result in several individual diagnostic plots with ODS names: **CooksDPlot, ObservedByPredicted, QQPlot**, and so on.

2.4.3 *Multiple Plot Requests*

To request multiple plots the plot names are enclosed in parentheses, e.g.

```
plots=(diagnostics residuals)
```

If options are required for any of these plot requests they can be added as before, e.g.

```
plots=(diagnostics(unpack) residuals)
```

Global plot options are those that apply to all plot requests and are specified by including them in parentheses after plots, as in

plots(unpack)=(diagnostics residuals)

One global plot option that is occasionally useful is the only option, as in plots(only)=residuals. The effect of this is to suppress any default plots that may otherwise be produced and include only the residual plots in the output.

It is possible to combine both global plot options with options for specific plot requests.

Details of the global plot options that are available and of those for each type of plot are given under the syntax tab of the procedure help for the appropriate statement, which is usually the proc statement.

2.4.4 Proc Plm

One way of extending the graphics available from a statistical procedure is by using proc plm. This procedure does not directly analyse a dataset, but uses the results of an analysis conducted by another procedure to produce further results, including graphics. The effectplot statement within proc plm can be used to display the fit of complex models. In version 9.4 of SAS, only three other statistical procedures have an effectplot statement, genmod, logistic and orthoreg. For other procedures that can store their results in an appropriate item store, proc plm offers the possibility of displaying their results in the same way. An example of its use is given in Chapter 6.

2.5 Controlling ODS Graphics

2.5.1 ODS Graphics Statement

The ods graphics statement enables and disables ODS graphics (by ods graphics on; and ods graphics off;) but also has options for controlling the graphics output. The most important of these are shown next.

Option	Description	Default
height	height of the graph	480px
width	width of the graph	640px
outputfmt	the image format (jpeg, tiff, png, ps, svg, etc.)	png
imagename	the ods name of the graph	
reset	reset graphic options to their default	all
border	whether a border is drawn (on/off)	on
attrpriority	Order for cycling group attributes (auto/colour/none)	auto

The height and width of a graph can also be specified in the other size units such as inches (in) or centimetres (cm). The default value of outputfmt depends on the ODS destination but is png for most of the commonly used destinations; for the html and listing destinations gif, jpeg and svg are among the alternatives, with tiff also available for listing. When multiple runs of a graphics procedure such as proc sgplot are made a sequence of files will be generated with names sgplot.png, sgplot1.png, sgplot2.png etc. The reset=index option sets this sequence back to the beginning, so that the next run will overwrite sgplot.png and the sequence will begin again from there. The imagename option is useful for giving meaningful names to graph files, e.g. imagename='Figure 2'.

2.5.2 ODS Styles

Whereas the ODS destination determines the general format of the output, its appearance in terms of fonts, colours and other attributes depends on the ODS style in use. There are a number of built-in styles and each output destination has a default style optimized for that destination. The default style for html is called htmlblue and those for the listing and rtf destinations are called listing and rtf, respectively.

The names of other built-in styles can be listed by submitting:

```
proc template;
   list styles;
run;
```

Where the final output is to be in black and white, with greyscale fills or shading, the journal style is a good choice. The graphs in this book have been produced using a version of the theme style adapted to give greyscale output rather than colour.

To change from the default style for the ODS destination, use the style= option on the ods destination statement, for example

```
ods listing style=journal;
```

2.6 Controlling Labelling in Graphs

2.6.1 Variable Labels

Although variable names are often suitable for use in graphs, variable labels offer much more flexibility: they can contain spaces and other characters not permitted in names and can be much longer. A label statement used within a proc step provides a label that is used for the duration of the step and may be used either to provide a label for a variable that does not have one or to override an existing label for the purposes of the graph.

An SAS system option controls whether or not labels are used in the output. The default is for them to be used but if this has been switched off, their use can be restored by submitting the statement **options label;**.

2.6.2 SAS Formats

SAS formats provide variable values with more meaningful labels. In the example code for Figure 2.4 formats were used to group values as well as to provide labels.

2.6.3 Titles and Footnotes

When graphs are produced by **sgplot**, **sgpanel** and **sgscatter**, the title and footnote statements may be used as usual, including the options font, height, color, bold and italic.

However, when graphs are produced by the statistical procedures the title and footnote statements generally do not apply. One way to change these is to use the ODS graphics editor as described in Section 2.7.

2.6.4 Axis Labelling

While the text used to label axes can be controlled by the label statement, the appearance of the text is determined by the ODS style. In the case of **sgplot** and **sgpanel** the appearance can be controlled by the labelattrs option on the axis statements. For **sgplot** the axis statements are xaxis and yaxis (plus x2axis and y2axis for secondary axes in overlaid plots); for **sgpanel** they are colaxis and row-axis. The label attributes that can be specified are colour, family (i.e. font type), size, style (italic/normal) and weight (bold/normal) as in the following example:

labelattrs=(color=red family=arial size=10 style=italic weight=bold)

2.7 ODS Graphics Editor

Changes and enhancements to an ODS graph can be made interactively using the ODS graphics editor.

To do this an editable version of the graph must be created, which is done by setting the **sge=on** option on the **ods** *destination* statement before creating the graph, as in the following example.

```
ods listing sge=on;
proc sgplot data=bodycomp;
  scatter y=pctfat x=age;
  label pctfat="Fat as a % of body mass";
run;
```

Having done this two versions of the plot will be visible in the Results window as shown in Figure 2.9. The second of these is an editable version. There will also be a corresponding file with the extension .sge. Double clicking on the entry in the results window will invoke the ODS graphics editor, which will appear similar to that shown in Figure 2.10.

The name of the graphics file and some properties of the graph appear in the title. Many of the menu items and buttons are standard to Windows. Hovering the cursor over the buttons will give a hint as to their use. Those in the second row, reading from right to left are to insert: a marker, an image, an ellipse, a rectangle, an arrow, a line, a piece of text. In Figure 2.10 we have inserted an arrow, some text and an ellipse.

Clicking on an axis label allows the text to be reformatted, while double clicking allows the text itself to be changed. Double clicking on the graph area allows some general properties of the graph to be changed.

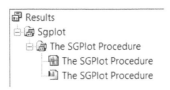

Figure 2.9   Segment of results window.

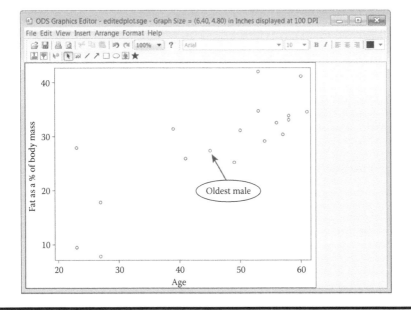

Figure 2.10   ODS graphics editor window.

Once the editing has been done, the file can be saved either as a .sge file or in the .png format. If another output format is required **proc sgrender** can be used to produce it as in the following example:

```
ods graphics on/outputfmt=jpeg;
proc sgrender sge="editedplot.sge";
run;
```

The ODS graphics editor can also be used in the same way to edit the graphs produced by statistical procedures. This is a big advantage as alternative methods of editing those graphs are generally much more complex.

Chapter 3

Graphs for Displaying the Characteristics of Univariate Data: Horse Racing, Mortality Rates, Forearm Lengths, Survival Times and Geyser Eruptions

3.1 Introduction

In this chapter we shall look at a variety of commonly used graphics for displaying the essential features, for example, the distribution, of a single variable (categorical or continuous) observed or measured on a sample of individuals or other entities of interest. We shall consider the following five sets of data:

> *Horse racing*—The file racestalls.dat contains a single variable giving the starting stall of the winners in 144 horse races held in the United States; all 144 races

took place on a circular track and all relate to eight stalls. Starting stall one is closest to the rail on the inside of the track. Interest here lies in assessing whether the chances of a horse winning a race are affected by its position in the starting lineup.

Mortality rates—The data shown in Table 3.1 are the under age 5 mortality rates per 1000 live births for a number of different countries (the data are from the United Nations Children's Fund 1992 publication, *The State of the World's Children*). The data are in the file **infmort.dat**. A graphic that

Table 3.1 Under 5 Mortality Rate per 1000 Live Births

Rwanda	198
Sudan	172
Laos	152
Lesotho	129
Guatemala	94
Turkey	80
Vietnam	65
Oman	49
Costa Rica	22
Belgium	9
Canada	9
Bolivia	160
Peru	116
Brazil	83
Argentina	35
Yemen	187
Saudi Arabia	91
Syria	59
Jordan	52
Israel	11

allows easy comparison of the mortality rates of the different countries is the goal here.

Forearm lengths—The data in Table 3.2 show the forearm lengths (in inches) of 140 adult males collected in an investigation of inheritance (see Pearson and Lee, 1903). The data are in a file **forearms.dat**. Here graphics for displaying the distribution of forearm lengths are required.

Survival times—The data shown in Table 3.3 are the survival times (in days) for patients suffering from gastric cancer, that is, the time from diagnosis until death. The data are in a file **cancer.dat**. Again displaying the distribution of the survival times graphically is the aim.

Geyser eruptions—For the final example in this chapter we will consider some data collected on the Old Faithful geyser. Geysers are natural fountains that shoot up into the air, at more or less regular intervals, a column of heated water and steam. Old Faithful is one such geyser and is the most popular attraction of Yellowstone National Park in the United States. Old Faithful can vary in height from 100 to 180 feet with an average near 130 to 140 feet. Eruptions normally last between 1.5 and 5 minutes. From August 1 to August 15, 1985, Old Faithful was observed and the waiting times

Table 3.2   Forearm Lengths (Inches) of 140 Adult Males

17.3	18.4	20.9	16.8	18.7	20.5	17.9	20.4	18.3	20.5
19.0	17.5	18.1	17.1	18.8	20.0	19.1	19.1	17.9	18.3
18.2	18.9	19.4	18.9	19.4	20.8	17.3	18.5	18.3	19.4
19.0	19.0	20.5	19.7	18.5	17.7	19.4	18.3	19.6	21.4
19.0	20.5	20.4	19.7	18.6	19.9	18.3	19.8	19.6	19.0
20.4	17.3	16.1	19.2	19.6	18.8	19.3	19.1	21.0	18.6
18.3	18.3	18.7	20.6	18.5	16.4	17.2	17.5	18.0	19.5
19.9	18.4	18.8	20.1	20.0	18.5	17.5	18.5	17.9	17.4
18.7	18.6	17.3	18.8	17.8	19.0	19.6	19.3	18.1	18.5
20.9	19.8	18.1	17.1	19.8	20.6	17.6	19.1	19.5	18.4
17.7	20.2	19.9	18.6	16.6	19.2	20.0	17.4	17.1	18.3
19.1	18.5	19.6	18.0	19.4	17.1	19.9	16.3	18.9	20.7
19.7	18.5	18.4	18.7	19.3	16.3	16.9	18.2	18.5	19.3
18.1	18.0	19.5	20.3	20.1	17.2	19.5	18.8	19.2	17.7

Table 3.3 Survival Time (Days) for Patients Suffering from Gastric Cancer

17	185	542
42	193	567
44	195	577
48	197	580
60	208	795
72	234	855
74	235	1174
95	254	1214
103	307	1232
108	315	1366
122	401	1455
144	445	1585
167	464	1622
170	484	1626
183	528	1736

between successive eruptions noted. There were 300 eruptions observed, so 299 waiting times (minutes) were recorded and it is these that are given in Table 3.4. The data are in a file **geyser.dat**. Interest here lies in using some suitable graphic for the data which might be helpful in suggesting a suitable model for the data.

3.2 Pie Chart, Bar Chart and Dot Plot Displays

Since a categorical variable merely divides the observations into groups, the main point of a graph for such a variable is to show how evenly, or unevenly, this is done. Newspapers, television and the media in general are very fond of displaying such data in the form of a *pie chart* or *bar chart*; in the former the sections of a circle have areas proportional to the observed percentages and in the latter the percentages are represented by rectangles of appropriate size placed along, commonly, a horizontal axis. Often this is for exploratory or descriptive purposes, but occasionally there will be a prior hypothesis about what form the distribution over the categories should take; the horse racing data is such an example.

Table 3.4   Waiting Times (Minutes) between Eruptions of Old Faithful

79	54	74	62	85	55	88	85	51	85	54	84	78	47	83	52	62	84	52	79	51	47	78	69	74
83	55	76	78	79	73	77	66	80	74	52	48	80	59	90	80	58	84	58	73	83	64	53	82	59
75	90	54	80	54	83	71	64	77	81	59	84	48	82	60	92	78	78	65	73	82	56	79	71	62
76	60	78	76	83	75	82	70	65	73	88	76	80	48	86	60	90	50	78	63	72	84	75	51	82
62	88	49	83	81	47	84	52	86	81	75	59	89	79	59	81	50	85	59	87	53	69	77	56	88
81	45	82	55	90	45	83	56	89	46	82	51	86	53	79	81	60	82	77	76	59	80	49	96	53
77	77	65	81	71	70	81	93	53	89	45	86	58	78	66	76	63	88	52	93	49	57	77	68	81
81	73	50	85	74	55	77	83	83	51	78	84	46	83	55	81	57	76	84	77	81	87	77	51	78
60	82	91	53	78	46	77	84	49	83	71	80	49	75	64	76	53	94	55	76	50	82	54	75	78
79	78	78	70	79	70	54	86	50	90	54	54	77	79	64	75	47	86	63	85	82	57	82	67	74
54	83	73	73	88	80	71	83	56	79	78	84	58	83	43	60	75	81	46	90	46	74			

3.2.1 Horse Racing

We begin by displaying the number of winners from each starting stall as a bar chart, using the **vbar** statement with **proc sgplot** as follows:

```
data racestalls;
   infile 'c:\HoSGuS\data\racestalls.dat';
   input stall_number;
   run;
proc sgplot data=racestalls;
   vbar stall_number;
run;
```

The **vbar** plot statement requires only the categorical variable, **stall_number**, to be specified. The default is to produce a vertical bar chart where the height of the bars represents the frequency of the observations in each category. The result is shown in Figure 3.1.

Figure 3.1 gives the impression that the chances of winning are indeed affected by the starting line up. There is perhaps a suggestion that horses starting in the higher numbered stalls are less likely to win. Under the null hypothesis that the probability of winning is the same for each starting position the expected number of winners from each starting position in 144 races is 18. We can put a reference line in the chart showing this, by adding a **refline** statement to the previous **proc sgplot** step.

```
proc sgplot data=racestalls;
   vbar stall_number;
   refline 18;
run;
```

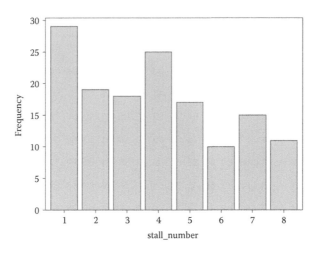

Figure 3.1   Vertical bar chart of starting stall number of 144 winning horses.

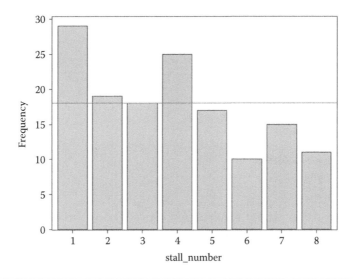

Figure 3.2   Vertical bar chart of starting stall number of 144 winning horses with reference line at the 'expected' value.

The **refline** statement adds one or more reference lines to a graph at the values specified. Here we have specified just the one value. By default the reference line is on the *y*-axis but the **axis=x** option can be used to add reference line on the *x*-axis.

The result is shown Figure 3.2. This plot suggests that the most likely departures from the equal probability hypothesis occur in stalls one and six and perhaps also stall eight.

Proc sgplot could also have been used to display these data as a horizontal bar chart, using the hbar statement, or, using the dot statement, as a *dot plot* where a 'dot' is used to indicate the relevant value in a category rather than the height of a rectangle. But bar charts, vertical and horizontal, and dot plots can also be constructed using proc freq, which is the main procedure for analysing the frequency distributions for categorical variables. The plots can be produced as ODS graphs with the plots option, which in the case of proc freq is specified on the tables statement. For example, we can produce a frequency dot plot for the horse race data as follows:

```
proc freq data=racestalls;
    tables stall_number/plots=freqplot(type=dotplot);
run;
```

The result is shown in Figure 3.3. The default type for freqplot is barchart and the orientation can be controlled by using the orient= option with the values horizontal or vertical.

When the chisq (chi square) option is being used to test for departure from the null hypothesis of equal proportions in each category a deviation plot can be produced as follows:

```
proc freq data=racestalls;
    tables stall_number/chisq plots=(deviationplot);
run;
```

The result is shown in Figure 3.4. In this plot, the *y*-axis is centred on the estimated expected value under the null hypothesis and the bars show the relative deviation of each category's frequencies from the expected value. So stall

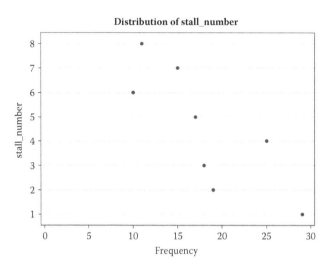

Figure 3.3   Dot plot of starting stall number of 144 winning horses.

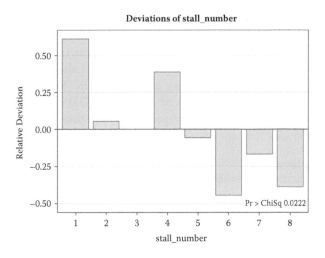

Figure 3.4   Deviation plot of starting stall number of 144 winning horses.

one with 29 winners is 61% above the expected value, whereas stall 6 with 10 winners is 44% below it. The *p*-value associated with the chi-square test shown on Figure 3.4 shows there is clear evidence of a departure from the equal probabilities hypothesis.

Perhaps the most frequently encountered plot for categorical variables is the *pie chart*. Neither proc sgplot nor proc freq produce pie charts, but the graphics template language may be used as follows:

```
proc template;
define statgraph pie;
   begingraph;
      layout region;
         piechart category=stall_number;
      endlayout;
   endgraph;
end;
run;

proc sgrender data=racestalls
              template=pie;
run;
```

The result is shown in Figure 3.5.

Even allowing for the fact that monochrome reproduction does not help, the pattern of wins is far less clear than it is in Figure 3.1. Nor is it at all clear how the pie chart could be enhanced to make the pattern clearer. Despite their widespread popularity, both the general and in particular the scientific use of pie charts has

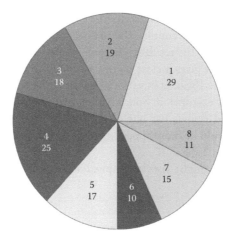

Figure 3.5   Pie chart of starting stall number of 144 winning horses.

been severely criticized. Tufte (1983), for example, comments that 'tables are preferable to graphics for many small datasets. A table is nearly always better than a dumb pie chart; the only worse design than a pie chart is several of them .... pie charts should never be used'. A similar lack of affection is shown by Bertin (1981), who declares that 'pie charts are completely useless', and by Wainer (1997) who claims that 'pie charts are the least useful of all graphical forms'.

3.2.2 Mortality Rates

One possibility for displaying the mortality data to allow easy comparison of the rates for different countries is an *ordered* dot plot, i.e. one where the 'dots' are arranged in order of the corresponding rates. Such a plot can be produced using the code

```
data infmort;
   infile 'c: \ hosgus \ data \ infmort.dat';
   input country $13. rate;
run;

proc sgplot data=infmort;
   dot country / freq=rate categoryorder=respdesc;
   xaxis label='Infant mortality per 1,000 live births';
run;
```

The syntax of the dot statement in **proc sgplot** is similar to that of the **vbar** statement earlier. In the previous example the data consisted of multiple observations for each value of the categorical variable, the starting stall position. In this case the data are already summarized in the form of rates. For the purpose of the plot we treat these rates as if they were counts of deaths and use the **freq=** option to specify that the variable **rate** contains the counts. The category axis is also ordered in descending order of the response. The resulting diagram is shown in Figure 3.6.

The ordered dot plot gives a clear picture of the mortality rates and how they increase from Canada to Rwanda in various regions of the world and the disparities in the same region, for example, in South America where Argentina has a relatively low mortality rate and Bolivia has one of the highest rates worldwide. For many people the dot plot would be more appealing to deal with than the simple listing of mortality rates given in Table 3.1.

The mortality rates might also be graphed as an ordered bar chart using the code

```
proc sgplot data=infmort;
   vbar country/freq=rate categoryorder=respasc;
   xaxis label='Infant mortality per 1,000 live births' fitpolicy=rotate;
run;
```

The result is shown in Figure 3.7. For the authors this figure does not give such an informative account of the data as Figure 3.6 possibly because its *data-ink ratio* (see Tufte, 1983) is lower than that of the dot plot. It also illustrates a common problem with vertical bar charts: there may not be enough space for the category labels, although the **fitpolicy=rotate** option does a relatively good job of dealing with the problem.

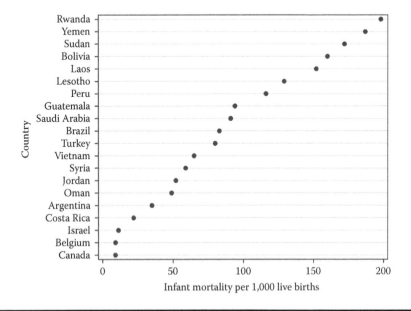

Figure 3.6   Dot plot for infant mortality data.

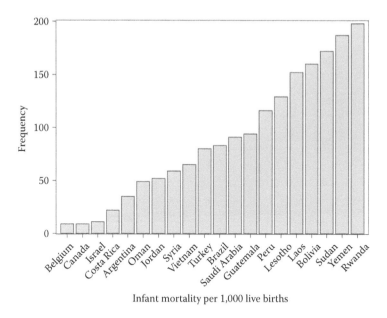

Figure 3.7 **Ordered bar chart for the infant mortality data.**

3.3 Histogram, Stem-and-Leaf and Boxplot Displays

3.3.1 Forearm Lengths

To investigate the distribution of these measurements we might begin by constructing a *histogram* in which the data have been grouped into a number of intervals of forearm length, and the number or frequency of observations in each interval is represented by the height of a rectangle positioned on the appropriate interval and with width equal to the width of the interval.

The histogram for the data in Table 3.2 can be produced using the following SAS code

```
data forearms;
infile 'c: \ hosgus \ data \ forearm.dat';
input length @@;
run;

proc sgplot data=forearms;
   histogram length;
run;
```

The resulting plot is shown in Figure 3.8. The histogram, which is essentially a crude estimate of the probability density function of a variable, suggests that the density of forearm length is symmetric and unimodal, which poses

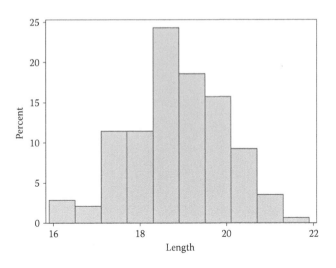

Figure 3.8   Histogram of forearm lengths.

the obvious question: Do these data follow a normal distribution? One way of informally addressing this question is to fit a normal distribution to the data and display the fitted normal on the observed histogram. This can be done by adding a density statement after the histogram statement.

```
proc sgplot data=forearms;
    histogram length;
    density length;
run;
```

The resulting plot is shown in Figure 3.9 and suggests that a normal distribution is a pretty safe assumption for these data. But we may want to assess the assumption somewhat more formally and this we can do by looking at a *normal probability plot* of the data. Such plots involve ordering the observations as say, $x_{(1)}, x_{(2)}, ..., x_{(n)}$ and then plotting them against the quantiles of a standard normal distribution, $\Phi^{-1}(p_i)$ where usually $p_i = (i - 0.5)/n$ and $\Phi(x) = \int_{-\infty}^{x} \frac{1}{\sqrt{2\pi}} e^{-\frac{1}{2}u^2} du$.

Proc univariate is the main procedure for examining the distributions of continuous variables and probability plots are amongst the ODS graphics that it can produce. To do this the probplot statement is used with the normal option and suboptions, which specify that the mean (mu) and standard deviation (sigma) of the underlying distribution are to be estimated from the data.

```
proc univariate data=forearms;
    var length;
    probplot length/normal(mu=est sigma=est);
run;
```

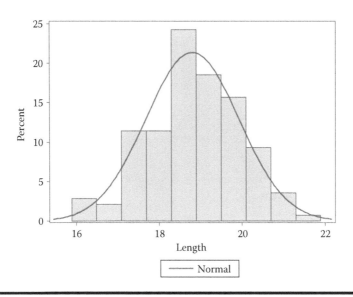

Figure 3.9   Histogram of forearm lengths and fitted normal distribution.

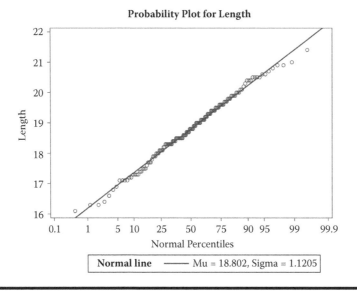

Figure 3.10   Normal probability plot of the forearm length data.

This gives Figure 3.10; there is little deviation from linearity confirming that the distribution of forearm lengths is very likely normal.

So a normal distribution appears suitable for the forearm data and proc univariate can also produce histograms and fit a number of other parametric distributions to these including: beta, gamma, exponential, lognormal and Weibull distributions.

But for many datasets it may be necessary to estimate the probability density function of a variable *without* assuming a particular parametric form. Perhaps the most common class of density estimators is the *kernel estimators* which are of the form

$$\hat{f}(x) = \frac{1}{nh} \sum_{i=1}^{n} K\left(\frac{x - X_i}{h}\right) \tag{3.1}$$

where h is known as *window width* or *bandwidth* and K is known as the *kernel function*, and is such that

$$\int_{-\infty}^{\infty} K(u)\, du = 1 \tag{3.2}$$

Essentially, such kernel estimators sum a series of 'bumps' placed at each of the observations. The kernel function determines the shape of the bumps while h determines their width. Three widely used kernel functions are

1. Gaussian

$$K(x) = \frac{1}{\sqrt{2\pi}} e^{-x^2/2} \tag{3.3}$$

2. Triangular

$$K(x) = 1 - |x|, \; |x| < 1 \tag{3.4}$$

3. Quadratic

$$K(x) = \frac{3}{4}(1 - x^2), \; |x| < 1 \tag{3.5}$$

A graphical representation of each kernel function is shown in Figure 3.11.

Figure 3.12 shows how the Gaussian kernel estimator works for an artificial dataset of observations, 0, 1, 1.1, 1.5, 1.9, 2.8, 2.9, 3.5.

In general, the choice of the shape of the kernel function is not usually of great importance. In contrast, the choice of bandwidth can be critical. There are situations in which it is satisfactory to choose the bandwidth relatively subjectively to achieve a 'smooth' estimate. More formal methods are however available. For details, see Silverman (1986).

To illustrate using kernel density estimates we can find the estimated probability density function of the forearm data using each of the three kernels defined

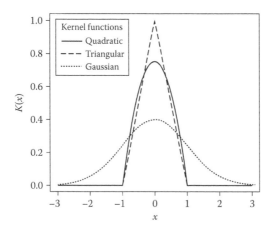

Figure 3.11   Three commonly used kernel functions.

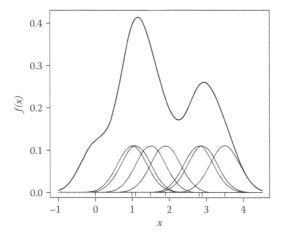

Figure 3.12   Kernel estimate showing the contributions of Gaussian (normal) kernels evaluated for the individual observations with bandwidth $h = 0.4$.

earlier and display the estimated densities on the observed histogram using the following SAS code:

```
proc sgplot data=forearms;
    histogram length;
    density length;
    density length/type=kernel(weight=normal) legendlabel='normal kernel';
    density length/type=kernel(weight=quadratic) legendlabel='quadratic';
    density length/type=kernel(weight=triangular)legendlabel='triangular';
run;
```

The first **density** statement fits a normal curve, which is the default, and the following three fit the different types of kernel density.

The result is shown in Figure 3.13; here the fitted normal and the three kernel estimated density functions are very similar confirming that the normal would be a suitable choice for this data. Note the use of the **legendlabel** option to distinguish the three kernel curves in the legend. Without this the three would all be labelled 'kernel'.

An alternative to the histogram as a way of displaying the fit of the data to a given distribution is a *hanging histogram*, which for the forearm data can be produced as follows:

```
proc univariate data=forearms;
  var length;
  histogram length/normal hang vref=0;
run;
```

In Figure 3.14 the bars of the histogram are 'hung' from the fitted distribution and the closer they are to the zero reference line the better the fit. These can be particularly useful for examining highly skewed datasets where detail of the fit at the peak may otherwise be difficult to assess.

A plot that has similar aims as the histogram in displaying the distribution of a variable is the *stem-and-leaf plot* but having the advantage that more of the data is

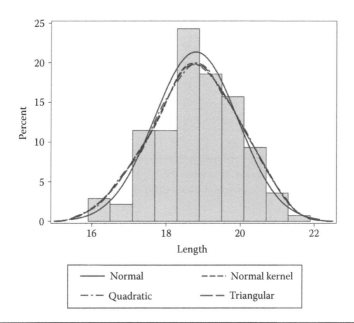

Figure 3.13   Histogram of forearm lengths showing fitted normal and three kernel estimates of the density function.

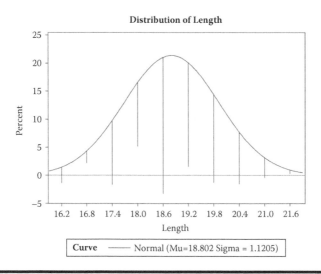

Figure 3.14   Hanging histogram for the forearm length data.

retained in the display. The stem-and-leaf plot for the forearm data is produced as follows (we shall explain the boxplot part of this plot directly):

```
proc univariate data=forearms plot;
  var length;
run;
```

```
    Stem Leaf                              #        Boxplot
    21 04                                  2          |
    20 5555667899                         10          |
    20 0001123444                         10          |
    19 5555666667778889999                19      +-----+
    19 00000011111222333344444            23      |     |
    18 55555555556666777788888999         25      *--+--*
    18 000111122333333334444              21      +-----+
    17 55567778999                        11          |
    17 111122333344                       12          |
    16 689                                 3          |
    16 1334                                4          |
         ----+----+----+----+----+
```

In the stem-and-leaf plot the shape of the variable's distribution is shown as it is in the histogram, but the variable values themselves are now also conveniently displayed, which can often be quite useful.

A further way to display the characteristics of a continuous variable is the *box-and-whisker plot* more usually referred to as a *boxplot*. This type of plot is based on the *five-number summary* of the data in which the five numbers are the minimum, the lower quartile (LQ), the median, the upper quartile (UQ) and the maximum.

The distance between the upper and lower quartiles, the *interquartile range* (IQR), is a measure of the *spread* of a variable's distribution. The relative distances from the median of the upper and lower quartiles give information about the *shape* of a variable's distribution; for example, if one distance is much greater than the other, the distribution is skewed. In addition, the median and the upper and lower quartiles can be used to define arbitrary but often useful limits, L and U, that maybe helpful in identifying possible outliers. The two limits are calculated as follows:

$$U = UQ + 1.5IQR$$

$$L = LQ - 1.5IQR$$

Observations outside the limits can be regarded as potential outliers (they are sometimes referred to specifically as *outside observations*) and such observations may merit careful attention before undertaking any analysis of a dataset because there is always the possibility that they can have undue influence on the results of the analysis.

The construction of a boxplot from the five-number summary is shown in Figure 3.15.

The boxplot of the forearms length data can be constructed using

```
proc sgplot data=forearms;
   vbox length;
run;
```

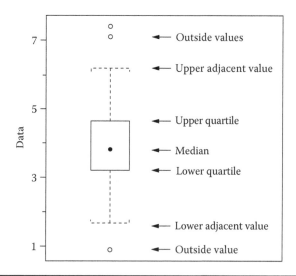

Figure 3.15 **The construction of a boxplot. (From B. S. Everitt, 2010, *Multivariable Modeling and Multivariate Analysis for the Behavioral Sciences*, CRC Press.)**

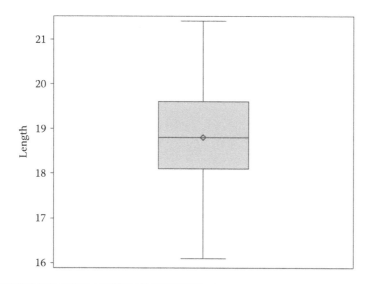

Figure 3.16   Boxplot of forearm lengths data.

The boxplot is shown in Figure 3.16. The default in SAS is for the median of the data to be represented by a line across the box and the mean by a diamond. Figure 3.16 demonstrates the symmetry of the data – the median line is in the centre of the box and the mean coincides with it. There are no unusually large or small observations in the data.

3.3.2 Survival Times

Moving on to the survival times data in Table 3.4 we can construct the relevant boxplot using the following code:

```
data cancer;
   infile 'c: \ hosgus \ data \ gastricca.dat';
   input days;
   id=_n_;
run;

proc sgplot data=cancer;
   vbox days / datalabel=id nocaps;
run;
```

The result is Figure 3.17. In this example, we choose to have the whiskers without caps at their ends and to label the outliers with their id number.

The boxplot in Figure 3.17 clearly indicates that the distribution of survival times is positively skewed—the median line is towards the bottom end of the box and

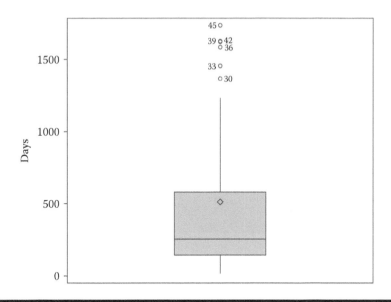

Figure 3.17   Boxplot of gastric cancer survival times.

clearly separated from the mean. There are also several outliers in the data. A possible distribution for such skewed data is the exponential and we can investigate whether or not this is reasonable for these survival times by using proc univariate to produce a hanging histogram of the data with an exponential distribution curve.

```
proc univariate data=cancer;
  var days;
  histogram days / exponential hang vref=0 nmidpoints=15;
run;
```

The resulting plot, shown in Figure 3.18, suggests that these data might have an exponential distribution, but once again we can assess this more formally with a probability plot suitable for the exponential, as follows:

```
proc univariate data=cancer;
  var days;
  probplot days / exponential(theta=est sigma=est);
run;
```

The resulting plot is shown in Figure 3.19. There appears to be some departure from the linearity required for an exponential to be convincingly assumed for these data.

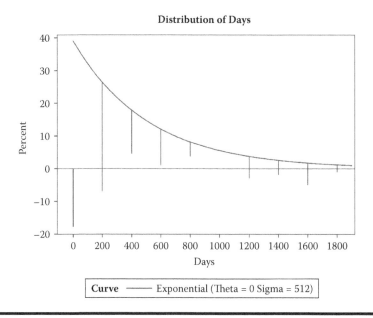

Figure 3.18 **Hanging histogram with fitted exponential distribution of the gastric cancer survival times.**

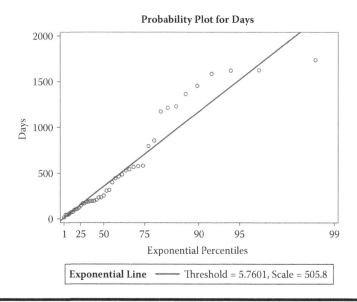

Figure 3.19 **Exponential probability plot for the cancer survival times.**

3.3.3 Geyser Eruptions

For the geyser eruptions data, we begin by constructing a histogram showing a fitted normal and a kernel estimate of the density function by using the following code:

```
data geyser;
    infile 'c:\hosgus\data\geyser.dat';
    input wait @@;
run;

proc sgplot data=geyser;
    histogram wait;
    density wait;
    density wait/type=kernel;
run;
```

The result is shown in Figure 3.20. Clearly a normal distribution is not suitable for these data; the kernel estimate strongly suggests that the distribution of waiting times is *bimodal*.

A possible model for a bimodal distribution of waiting times is a *two-component normal mixture* given by

$$f(t) = p_1 \frac{1}{\sigma_1 \sqrt{2\pi}} \exp\left[-\frac{1}{2}\left(\frac{t-\mu_1}{\sigma_1}\right)^2\right] + (1-p_1)\frac{1}{\sigma_2 \sqrt{2\pi}} \exp\left[-\frac{1}{2}\left(\frac{t-\mu_2}{\sigma_2}\right)^2\right] \quad (3.6)$$

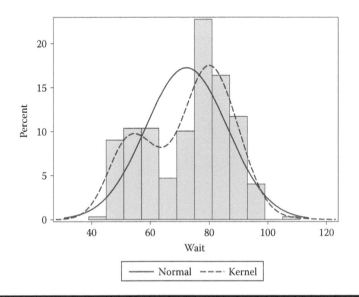

Figure 3.20 **Histogram of waiting times between eruptions of Old Faithful showing a fitted normal and a kernel estimate of the density function.**

For details of such mixture distributions see Everitt and Hand (1981), or Everitt et al. (2011); both of these references show how the five parameters of the distribution, p_1, μ_1, μ_2, σ_1, σ_2 can be estimated by maximum likelihood. For interest we will show how to do this estimation with SAS and also show how the fitted distribution can be shown on the histogram of the data. The following code does both:

```
proc fmm data=geyser;
  model wait=/dist=normal k=2 parms(55., 80.);
run;
```

Proc fmm fits models to data where the outcome has a mixture distribution but can also be used more simply to estimate the parameters of the distribution as we do here. The **model** statement specifies **wait** as the outcome but no predictors are given. The options on the **model** statement specify that the component distributions are normal; that there are two (**k=2**) components; and gives starting values for the means and variances of the component distributions via the **parms** option. Here we just give starting values for the means, having estimated them by eye from Figure 3.20; the dots could be replaced with starting values for the variances, if necessary. The fitted mixture is shown on the histogram of the data in Figure 3.21 with the estimated mean and variance of the fitted normal distributions given in the inset. The mixing proportions were estimated at 31%:69%. The mixture distribution is clearly an excellent fit.

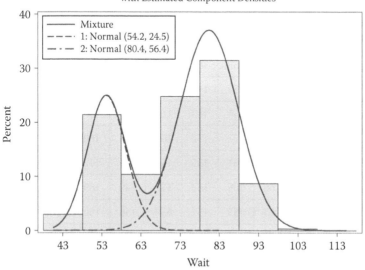

Distribution and Estimated Density for Wait
with Estimated Component Densities

Figure 3.21 Density plot from a two-component mixture model of the Old Faithful geyser eruption times.

3.4 Summary

The plots described in this chapter are often very useful for displaying the characteristics of a single variable; often, but not in every case. Pie charts, for example, although widely used are rarely better at communicating the values associated with the categories of a categorical variable than the table of numbers from which the chart is produced. To a lesser extent the same comment can sometimes be applied to bar charts. Histograms, which are widely used to display the distributions of continuous variables, can be greatly improved by enhancing them with a fitted distribution, for example, a normal distribution or one estimated using a kernel density estimator.

Exercises

3.1. The data in Table 3.5 (elderly.dat) are the heights in centimetres of a sample of 351 women randomly selected from the community in a study of osteoporosis. Investigate these data using suitable graphical displays.

Table 3.5   Heights in Centimetres of a Sample of 351 Women with Osteoporosis

156	163	169	161	154	156	163	164	156	166	177	158
150	164	159	157	166	163	153	161	170	159	170	157
156	156	153	178	161	164	158	158	162	160	150	162
155	161	158	163	158	162	163	152	173	159	154	155
164	163	164	157	152	154	173	154	162	163	163	165
160	162	155	160	151	163	160	165	166	178	153	160
156	151	165	169	157	152	164	166	160	165	163	158
153	162	163	162	164	155	155	161	162	156	169	159
159	159	158	160	165	152	157	149	169	154	146	156
157	163	166	165	155	151	157	156	160	170	158	165
167	162	153	156	163	157	147	163	161	161	153	155
166	159	157	152	159	166	160	157	153	159	156	152
151	171	162	158	152	157	162	168	155	155	155	161
157	158	153	155	161	160	160	170	163	153	159	169

Table 3.5 (*Continued*) Heights in Centimetres of a Sample of 351 Women with Osteoporosis

155	161	156	153	156	158	164	160	157	158	157	156
160	161	167	162	158	163	147	153	155	159	156	161
158	164	163	155	155	158	165	176	158	155	150	154
164	145	153	169	160	159	159	163	148	171	158	158
157	158	168	161	165	167	158	158	161	160	163	163
169	163	164	150	154	165	158	161	156	171	163	170
154	158	162	164	158	165	158	156	162	160	164	165
157	167	142	166	163	163	151	163	153	157	159	152
169	154	155	167	164	170	174	155	157	170	159	170
155	168	152	165	158	162	173	154	167	158	159	152
158	167	164	170	164	166	170	160	148	168	151	153
150	165	165	147	162	165	158	145	150	164	161	157
163	166	162	163	160	162	153	168	163	160	165	156
158	155	168	160	153	163	161	145	161	166	154	147
161	155	158	161	163	157	156	152	156	165	159	170
160	152	153									

3.2. A sequence of 300 pseudo-random digits was generated on a small calculator. The data in Table 3.6 give the number of times each of the digits 0, 1, …, 9 occurred. Do these data look like a sample from a discrete uniform distribution on 1, 2, …, 9? Investigate this question using whichever graphics you think will be most helpful.

Table 3.6 Distribution of Pseudo-Random Digits

Digit	0	1	2	3	4	5	6	7	8	9
Frequency	25	28	29	35	35	31	27	33	32	25

3.3. The data in Table 3.7 (velocities.dat) are the velocities of 82 galaxies from six well-separated conic sections of space. The data are intended to shed light on whether or not the observable universe contains superclusters of galaxies surrounded by large voids. The evidence for superclusters would be the multimodality of the distribution of velocities. Investigate by graphical means whether or not the velocities in Table 3.7 give any evidence for the superclusters idea.

Table 3.7   Velocities of 82 Galaxies from Six Well-Separated Conic Sections of Space

9172	9558	10406	18419	18927	19330	19440	19541	19846	19914	19989
20179	20221	20795	20875	21492	21921	22209	22314	22746	22914	23263
23542	23711	24289	24990	26995	34279	9350	9775	16084	18552	19052
19343	19473	19547	19856	19918	20166	20196	20415	20821	20986	21701
21960	22242	22374	22747	23206	23484	23666	24129	24366	25633	32065
9483	10227	16170	18600	19070	19349	19529	19663	19863	19973	20175
10215	20629	20846	21137	21814	22185	22249	22495	22888	23241	23538
23706	24285	24717	26960	32789						

Chapter 4

Graphs for Displaying Cross-Classified Categorical Data: Germinating Seeds, Piston Rings, Hodgkin's Disease and European Stereotypes

4.1 Introduction

In this chapter we will look at graphics that might be helpful in understanding data arising from counts observed in the cells of a cross-classification of two or more categorical variables. The following four sets of data will be used.

Germinating seeds—The data given in Table 4.1 arise from an experiment to study the effect of different amounts of water on the germination of seeds. For each amount of water, four identical boxes were sown with 100 seeds each and the number of seeds having germinated after two weeks was recorded. The experiment was repeated with the boxes covered to slow evaporation. There were six levels of watering coded 1 to 6 with higher codes corresponding to more water.

Table 4.1   Germinating Seeds

		Water Amount					
Covered	Box	1	2	3	4	5	6
No	1	22	41	66	82	79	0
	2	25	46	72	73	68	0
	3	27	59	51	73	74	0
	4	23	38	78	84	70	0
Yes	1	45	65	81	55	31	0
	2	41	80	73	51	36	0
	3	42	79	74	40	45	0
	4	43	77	76	62	*	0

*No result was recorded for this box.

Table 4.2   Data on Piston Ring Failures

Compressor No.	North	Centre	South	Total
1	17	17	12	46
2	11	9	13	33
3	11	8	19	38
4	14	7	28	49
Total	53	41	72	166

Piston rings—The data in Table 4.2 give the number of piston ring failures in each of three legs of four steam-driven compressors located in the same building (Haberman, 1973). The compressors have identical designs and are orientated in the same way. The question of interest is whether the two categorical variables, compressor and leg, are independent.

Hodgkin's disease—The data in Table 4.3 were recorded during a study of Hodgkin's disease, a cancer of the lymph nodes (see Hancock et al., 1979). Each of 538 patients with the disease was classified by histological type and by their response to treatment three months after it had begun. The histological types are LP = lymphocyte predominance, NS = nodular sclerosis, MC = mixed cellularity and LD = lymphocyte depletion. The question of

Table 4.3   Hodgkin's Disease

Histological Type	Response			
	Positive	*Partial*	*None*	*Total*
LP	74	18	12	104
NS	68	16	12	96
MC	154	54	58	266
LD	18	10	44	72
Total	314	98	126	538

Table 4.4   Perceived Characteristics by UK Nationals of Nationals from Other Countries in the European Union

Country	Characteristic												
	1	*2*	*3*	*4*	*5*	*6*	*7*	*8*	*9*	*10*	*11*	*12*	*13*
France	37	29	21	19	10	10	8	8	6	6	5	2	1
Spain	7	14	8	9	27	7	3	7	3	23	12	1	3
Italy	30	12	19	10	20	7	12	6	5	13	10	1	2
UK	9	14	4	6	27	12	2	13	26	16	29	6	25
Ireland	1	7	1	16	30	3	10	9	5	11	22	2	27
Holland	5	4	2	2	15	2	0	13	24	1	28	4	6
Germany	4	48	1	12	3	9	2	11	41	1	38	8	8

Note: Characteristics—1, stylish; 2, arrogant; 3, sexy; 4, devious; 5, easy-going; 6, greedy; 7, cowardly; 8, boring; 9, efficient; 10, lazy; 11, hard-working; 12, clever; 13, courageous.

interest here is what, if any, is the relationship between histological type and response to treatment.

European stereotypes—The data in Table 4.4 were obtained by asking a large number of people in the United Kingdom which of 13 characteristics they would associate with the nationals of the United Kingdom's partner countries in the European Community. Entries in the table give the percentages of respondents agreeing that the nationals of a particular country possess the particular characteristic.

4.2 Bar Charts for Cross-Classified Categorical Data

4.2.1 Seed Germination

A useful bar chart for the seed germination data can be produced using the following code:

```
data seeds;
  do covered=0 to 1;
  do box=1 to 4;
  do water=1 to 6;
  input ngerm @@;
  output;
  end;
  end;
  end;
cards;
22 41 66 82 79 0
25 46 72 73 68 0
27 59 51 73 74 0
23 38 78 84 70 0
45 65 81 55 31 0
41 80 73 51 36 0
42 79 74 40 45 0
43 77 76 62 .  0
;

proc format;
   value yn 0='No' 1='Yes';
   value covered 0='Uncovered' 1='Covered';
run;
proc sgpanel data=seeds;
   panelby covered water/layout=lattice rows=6;
   vbar box/response=ngerm;
   format covered yn.;
   rowaxis label='Number germinated';
run;
```

The panelby statement uses the lattice layout whereby the rows and columns are formed from the two variables covered and water. The rows=6 option prevents the graph from being split into two parts.

The resulting graphic is shown in Figure 4.1. This clearly shows that for watering levels one and two the covered boxes produce more germination than the uncovered boxes. For watering level three the covered boxes give approximately the same amount of germination as the uncovered boxes and for watering levels four and five the uncovered boxes give a higher degree of germination than the covered boxes. For watering level six both uncovered and covered boxes gave no germination at all.

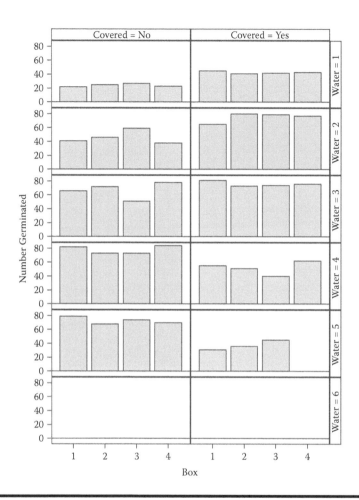

Figure 4.1   Bar chart for seed germination data.

Another useful graphic for the seed germination data can be produced for the data averaged over the four boxes using the following code:

```
proc sgpanel data=seeds noautolegend;
   panelby covered/novarname spacing=10;
   vbar water/response=ngerm stat=mean limits=both;
   format covered covered.;
   label water='Watering level';
   rowaxis label='Mean number germinated';
run;
```

For the **vbar** statement the default statistic for a response variable is the sum. Here we specify the mean and its confidence limits. The appearance is enhanced

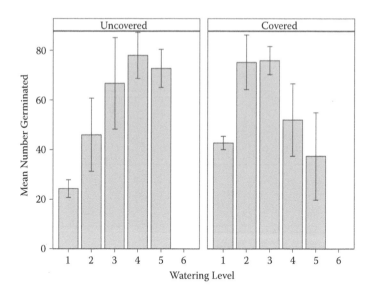

Figure 4.2   Bar chart for seed germination averaged over the four boxes.

by using a descriptive format for **covered** together with the **novarname** option on the **panelby** statement; adding some spacing between the panels and more explicit labeling of the axes with the **label** and **rowaxis** statement.

The resulting diagram is shown in Figure 4.2. This reinforces what is shown in Figure 4.1. Covered boxes lead to significantly more germination than uncovered boxes for watering levels one and two. For watering level three there is essentially no difference in the degree of germination in covered and uncovered boxes, and for watering levels four and five the germination is greater in uncovered boxes that in covered. For both uncovered and covered boxes the sixth level of watering stops germination all together.

4.2.2 Hodgkin's Disease

We can use these data to illustrate what is usually known as a *stacked bar chart* although the term *segmented bar chart* is also used. Here the horizontal axis represents one variable with each bar representing one of the categories of the variable. Each of these bars is then segmented according to the categories of the second variable. Each bar is then made up of smaller bars stacked on top of each other. To construct the required graphic we can use the following code:

```
data Hodgkins;
 input type$ v1-v3;
 response='Positive'; n=v1; output;
```

```
  response='Partial'; n=v2; output;
  response='None'; n=v3; output;
  drop v1-v3;
cards;
LP   74 18 12
NS   68 16 12
MC 154 54 58
LD   18 10 44
;

proc sgplot data=hodgkins;
   vbar type/group=response response=n;
   label type='Histological type';
   yaxis label='frequency';
run;
```

When the **vbar** plotting statement is used with a **group** variable, the default is to stack the bars. The alternative is a clustered bar chart with the **groupdisplay= cluster** option.

The resulting stacked bar chart is shown in Figure 4.3; this graphic clearly demonstrates the relatively poor response to treatment of those patients diagnosed as having lymphocyte depletion and the relatively good and almost equal response profile of patients in the lymphocyte predominance and nodular sclerosis diagnostic groups.

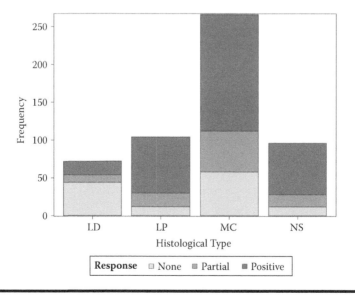

Figure 4.3   Stacked bar chart for Hodgkin's disease data.

4.3 Mosaic and Association Plots for Cross-Classified Data

Mosaic and *association plots* allow an investigator to examine the relationship between two categorical variables. The mosaic plot starts as a square of length one, which is first divided into horizontal bars whose widths are proportional to the percentages associated with the first categorical variable. Then each bar is split vertically into bars that are proportional to the *conditional percentages* of the second categorical variable within each category of the first variable.

An association plot gives a graphical bar chart like representation of essentially the differences between the observed frequencies in the data and the estimated frequencies under the hypothesis of the independence of compressor and leg; these differences are often called 'residuals'; see Equation 4.1 for the formal definition.

If we consider first the piston ring failure data, the independence of leg and compressor for the data would normally be addressed by the application of the familiar chi-square test of independence. If this is done the test statistics has a value of 1.72 with 6 degrees of freedom and an associated *p*-value of 0.068. It seems that the data provide no strong evidence of a departure from independence of compressor and leg but here we look at graphics for the data which may perhaps give more insight into the data.

Let's begin by constructing a mosaic plot for the piston ring failure data. The required code is

```
data pistons;
    input Compressor Leg $ n;
cards;
1 North    17
1 Centre   17
1 South    12
2 North    11
2 Centre    9
2 South    13
3 North    11
3 Centre    8
3 South    19
4 North    14
4 Centre    7
4 South    28
;
run;

proc freq data=pistons;
    tables compressor*leg/chisq out=tabout outexpect plot=mosaic;
    weight n;
run;
```

Proc freq is the main procedure used for analyzing cross-classified data and can be used to produce mosaic plots via the plot option on the tables statement. The out= and outexpect options are for later use. Since the variables compressor and leg were first defined (in the data step) with an initial capital letter, that is how they appear in the plot.

The resulting plot is given in Figure 4.4. (The small margins around each rectangle in the plot are inserted to make the graph easier to read.) This diagram clearly illustrates that the largest number of piston ring failures is in the South leg and particularly in the South leg of compressor four. Failures in the Centre and North legs are less frequent than in the South leg, and Compressor one suffers from the most failures in the Centre leg. Failures in the North leg appear to be approximately equally divided between the four compressors. The diagram might be very helpful in discussing the data with, say, the engineer responsible for the maintenance of the compressors.

We can construct an association plot for these data using the following code:

```
data resids;
  set tabout;
  residual=(count-expected)/sqrt(expected);
run;

proc sgpanel data=resids;
  panelby compressor/layout=rowlattice rows=4 noborder;
  vbar leg/response=residual;
  rowaxis label='Residual';
run;
```

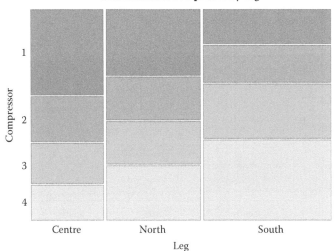

Figure 4.4   Mosaic plot of piston ring failure data.

In the proc freq step above, the out=tabout and outexpect options saved the frequency counts and expected values to the dataset tabout. These are then used to calculate residuals, which are then plotted by compressor and leg. The noborder option on the panelby statement gives a cleaner appearance to the graph.

The resulting plot is shown in Figure 4.5 and shows an interesting pattern with, in particular, compressors one and four having relatively large residuals for the Centre and the South legs but in different directions. Compressors two and three have relatively small residuals for all three legs. Perhaps the non-significant chi-square test of the independence of compressor and leg does not tell the whole story about these data?

Now we will move on to the Hodgkin's disease data and again produce both mosaic and association plots using the same code as used earlier but now with the

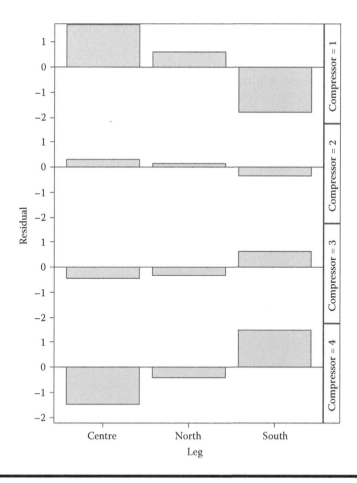

Figure 4.5   Association plot for the piston ring data.

Distribution of Type by Response

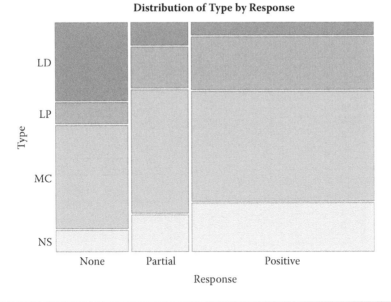

Figure 4.6 Mosaic plot for Hodgkin's disease data.

different dataset. The mosaic plot is shown in Figure 4.6 and the association plot in Figure 4.7. From Figure 4.6 we can see that there are more overall positive responses to treatment than partial or no responses. Patients with lymphocyte depletion are seen to have the worst response to treatment with only a small proportion of such patients having a positive response and a far larger proportion having no response. Having mixed cellularity also appears to be associated with relatively poor response to the treatment.

Moving on to Figure 4.7 we can see again the pattern that patients in the diagnostic category lymphocyte depletion have more people in the 'no response' category than would be expected and patients in the lymphocyte predominance have more in the 'positive' response category than would be expected if response was independent of diagnostic group.

4.4 Correspondence Analysis

Correspondence analysis is a technique for displaying the associations among a set of categorical variables in a type of scatterplot or map, thus allowing a visual examination of the structure or pattern of these associations. A correspondence analysis should ideally be seen as an extremely useful supplement to, rather than a replacement for, the more formal inferential procedures

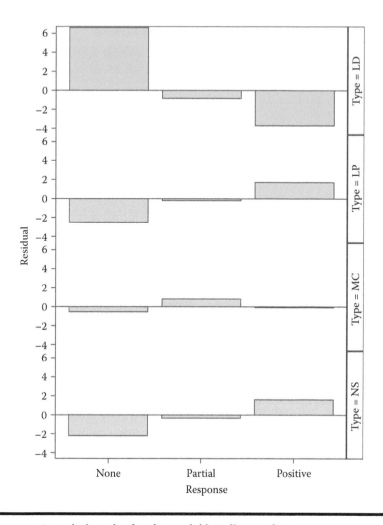

Figure 4.7   Association plot for the Hodgkin's disease data.

generally used with categorical data (see Chapters 3 and 8). The aim when using correspondence analysis is nicely summarized in the following quotation from Greenacre (1992):

> An important aspect of correspondence analysis which distinguishes it from more conventional statistical methods is that it is not a confirmatory technique, trying to prove a hypothesis, but rather an exploratory technique, trying to reveal the data content. One can say that it serves as a window onto the data, allowing researchers easier access to their numerical results and facilitating discussion of the data and possibly generating hypotheses which can be formally tested at a later stage.

Mathematically correspondence analysis can be regarded as either

■ a method for decomposing the chi-square statistic for a contingency table into components corresponding to different dimensions of the heterogeneity between its rows and columns, or
■ a method for simultaneously assigning a scale to rows and a separate scale to columns so as to maximize the correlation between the resulting pair of variables.

Quintessentially, however, correspondence analysis is a technique for displaying multivariate categorical data graphically, by deriving coordinate values to represent the categories of the variables involved, which may then be plotted to provide a 'picture' of the data.

In the case of two categorical variables forming a two-dimensional contingency table, the required coordinates are obtained from the *singular value decomposition* (Everitt and Dunn, 2001) of a matrix E with elements e_{ij} given by

$$e_{ij} = \frac{p_{ij} - p_{i+}p_{j+}}{\sqrt{p_{i+}p_{+j}}} \tag{4.1}$$

where $p_{ij} = n_{ij}/n$ with n_{ij} being the number of observations in the ijth cell of the contingency table and n the total number of observations; the total number of observations in row i is represented by n_{i+}, and the corresponding value for column j, n_{+j}. Finally $p_{i+} = n_{i+}/n$ and p_{+j}/n. The elements of E can be written in terms of the familiar 'observed' (O) and 'expected' (E) nomenclature used for contingency tables as

$$e_{ij} = \frac{1}{\sqrt{n}} \frac{O - E}{\sqrt{E}} \tag{4.2}$$

Written in this way it is clear that the terms are a form of residual from fitting the independence model to the data.

The singular value decomposition of E consists of finding matrices, U, V and Δ (diagonal) such that

$$E = U\Delta V' \tag{4.3}$$

where U contains the eigenvectors of EE' and V the eigenvectors of $E'E$. The diagonal matrix Δ contains the ranked singular values, δ_k so that δ_k^2 are the eigenvalues (in decreasing) order of either EE' or $E'E$.

The coordinate of the row ith category on the kth coordinate axis is given by $\delta_k u_{ik}/\sqrt{p_{i+}}$, and the coordinate of the jth column category on the same axis is given in $\delta_k v_{jk}/\sqrt{p_{+j}}$, where u_{ik}, $i = 1...r$, and v_{jk}, $j = 1...c$ are, respectively, the elements of the kth column of U and the kth column of V.

To represent the table fully requires at most $R = \min(r,c) - 1$ dimensions, where r and c are the number of rows and columns of the table. R is the rank of the matrix E. The eigenvalues, δ_k^2, are such that

$$\text{Trace}\ (EE') = \sum_{k=1}^{R} \delta_k^2 = \sum_{i=1}^{r} \sum_{j=1}^{c} e_{ij}^2 = \frac{X^2}{n} \tag{4.4}$$

where X^2 is the usual chi-square test statistic for independence. In the context of correspondence analysis, X^2/n is known as *inertia*. Correspondence analysis produces a graphical display of the contingency table from the columns of U and V, in most cases from the first two columns, u_1, u_2, v_1, v_2, of each, since these give the 'best' two-dimensional representation. It can be shown that the first two coordinates give the following approximation to the e_{ij}:

$$e_{ij} \approx u_{i1}v_{j1} + u_{i2}v_{j2} \tag{4.5}$$

so that a large positive residual corresponds to u_{ik} and v_{jk} for $k = 1$ or 2, being large and of the same sign. A large negative residual corresponds to u_{ik} and v_{jk}, being large and of opposite sign for each value of k. When u_{ik} and v_{jk} are small and their signs are not consistent for each k, the corresponding residual term will be small. The adequacy of the representation produced by the first two coordinates can be assessed informally by calculating the percentages of the inertia they account for, i.e.

$$\text{percentage inertia} = \frac{\delta_1^2 + \delta_2^2}{\sum_{k=1}^{R} \delta_k^2} \tag{4.6}$$

Values of 60% and over usually mean that the two-dimensional solution gives a reasonable account of the structure in the table.

(Displaying the categories of cross-classified data in this way involves the concept of *distance* between the percentage profiles of row and column categories. The distance measure used in a correspondence analysis is known as the *chi-square distance*. Details are given in Der and Everitt, 2009.)

In general what is of most interest in correspondence analysis solutions is the joint interpretation of the points representing the row and column categories in the two-dimensional map that results. It can be shown that row and column coordinates that are large and of the same sign correspond to a large positive 'residual' (in essence the difference between the observed frequency and that expected under independence) in the cross-classification. Row and column coordinates that are large but of opposite signs imply a cell in the cross-classification with a large negative residual. Finally, small coordinate values close to the origin correspond to small residuals.

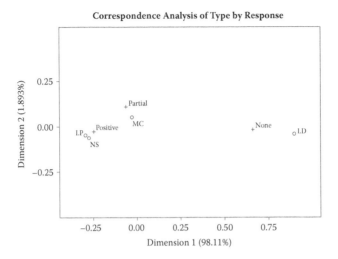

Figure 4.8   Correspondence analysis of Hodgkin's disease data.

4.4.1 Hodgkin's Disease

We can apply correspondence analysis to the data in Table 4.3 using the following code:

```
proc corresp data=hodgkins;
   tables type, response;
   weight n;
run;
```

The **tables** statement specifies the row variable, **type**, followed by a comma and the column variable, **response**. As the data already consist of counts, the **weight** statement specifies the variable containing the counts, **n**. The default number of dimensions for the analysis is two; other values can be specified via the **dimens=***n* option on the proc statement. Here we shall concentrate on the two-dimensional solution and the configuration plot, produced by default, given in Figure 4.8. Here we can see that the first dimension is dominant and the position of the points representing the histological type and response indicate clearly that a positive response is associated with lymphocyte predominance and nodular sclerosis, and no response is associated with lymphocyte depletion.

4.4.2 European Stereotypes

Correspondence analysis can be applied to the data in Table 4.4 using the following code:

```
data europeans;
   infile "c: \ hosgus \ data \ europeans.dat" expandtabs;
```

```
    input country $ c1-c13;
    label c1='stylish'
          c2='arrogant'
          c3='sexy'
          c4='devious'
          c5='easy-going'
          c6='greedy'
          c7='cowardly'
          c8='boring'
          c9='efficient'
          c10='lazy'
          c11='hard working'
          c12='clever'
          c13='courageous';
run;

proc corresp data=europeans out=coor;
  var c1-c13;
  id country;
run;
```

In this example, the data are already in the form of a table and the columns are specified on the var statement with the row identifiers given on the id statement. The resulting two-dimensional correspondence analysis solution is shown in Figure 4.9.

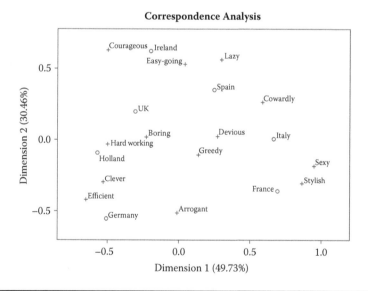

Figure 4.9   Correspondence analysis of the perceived characteristics of Europeans of different nationalities.

No doubt any French readers will be happy to find they are considered stylish and sexy but perhaps Italians will be less pleased with the tendency to be seen as devious and cowardly. Germans are, of course, well known to be efficient and clever, which can produce a perceived arrogance as shown in the plot. Finally we in the United Kingdom are judged little more than boring, which all of us know to be a mistake!

4.5 Summary

Graphical representations of cross-classified categorical data can be useful supplements to the usual statistical tests applied routinely to such data. Just as most researchers (at least the good ones) would not try to interpret a correlation coefficient and the result of a test for some hypothesis about the population value without looking at the associated scatterplot, it might be sensible not to simply take, say, the value of a chi-square test for the independence of two categorical variables at face value but instead interpret it along with one or other of the plots described in this chapter.

Exercises

4.1. The data in Table 4.5 (taken from Neyzi et al., 1975) were collected to investigate the effect of socio-economic class on physical development. Breast sizes were classified on a scale of 1 (no development) to 5 (fully developed) and the socio-economic class of their parents were assessed on a scale of 1 to 4. Investigate these data using suitable graphics and compare your conclusions with the result of a chi-square test of the independence of breast development and socio-economic class.

Table 4.5 Breast Development in Turkish Girls

Socio-Economic Class of Parents	Breast Development				
	1	*2*	*3*	*4*	*5*
1	2	14	28	40	18
2	1	21	25	25	9
3	1	12	12	12	2
4	6	17	34	33	6

4.2. The data in Table 4.6 describe the spatial association between tree species in Lansing Wood, Michigan (from Digby and Kempton, 1987). The entries in the table are the number of times that each species occurs as the nearest neighbour of each other species. Use correspondence analysis to investigate the data and discuss your conclusions about the spatial association of different species of tree in this wood.

Table 4.6 Trees' Nearest Neighbours

	Number of Occurrences as Nearest Neighbour						
	Red Oak	White Oak	Black Oak	Hickory	Maple	Other	Trees Total
Red oak	104	59	14	95	64	10	346
White oak	62	138	20	117	95	16	448
Black oak	12	20	27	51	25	0	135
Hickory	105	108	48	355	71	16	703
Maple	74	70	21	79	242	28	514
Other	11	14	0	25	30	25	105

4.3. Table 4.7 shows a three-dimensional contingency table containing data on suicide behaviour. Using any of the graphic displays described in this chapter investigate how the cause of death from suicide is related to the age and the gender of the individual committing suicide.

Table 4.7 Suicide Behaviour by Age, Gender and Cause of Death

	Age	Poisoning	Gas	Hanging/ Drowning	Gun/ Knife	Jumping	Other
Male							
	10–40	398	121	455	155	55	124
	41–70	399	82	797	168	51	82
	>70	93	6	316	33	26	14
Female							
	10–40	259	15	95	14	40	38
	41–70	450	13	450	26	71	60
	>70	154	5	185	7	38	10

Chapter 5

Graphs for Use When Applying *t*-Tests and Analyses of Variance: Skulls, Cancer Survival Times and Effect of Smoking on Performance

5.1 Introduction

In this chapter we will concern ourselves with the following three sets of data:

Head breadths of skulls—In an anthropometric study, measurements of the maximum head breadth (measured in millimetres) were taken on 84 skulls of Etruscan males and for a sample of 70 modern Italian males. Full details are given in Barnicot and Brothwell (1959). The data are shown in Table 5.1. The data were collected to shed light on the origins of the Etruscan empire (which is, we are told, something of a mystery to anthropologists) and a particular question is whether Etruscans were native Italians or immigrants from elsewhere.

Survival times for cancer—Our second set of data is taken from Cameron and Pauling (1978) and involves the survival times (in days) of patients with

Table 5.1   Maximum Head Breadths (mm) for 84 Etruscan and 70 Italian Skulls

Etruscans											
141	148	132	138	154	142	150	146	155	158	150	140
147	148	144	150	149	145	149	158	143	141	144	144
126	140	144	142	141	140	145	135	147	146	141	136
140	146	142	137	148	154	137	139	143	140	131	143
141	149	148	135	148	152	143	144	141	143	147	146
150	132	142	142	143	153	149	146	149	138	142	149
142	137	134	144	146	147	140	142	140	137	152	145
Italians											
133	138	130	138	134	127	128	138	136	131	126	120
124	132	132	125	139	127	133	136	121	131	125	130
129	125	136	131	132	127	129	132	116	134	125	128
139	132	130	132	128	139	135	133	128	130	130	143
144	137	140	136	135	126	139	131	133	138	133	137
140	130	137	134	130	148	135	138	135	138		

advanced cancer of the stomach, bronchus, colon, ovary or breast when treated with ascorbate. The data are given in Table 5.2. Clearly the question of most interest is whether survival times differ with the organ affected by the cancer.

Effect of smoking on performance—A study reported in Spilich et al. (1992) investigated the effects of smoking on performance. Three tasks were used that differ in the level of cognitive processing that was needed to perform them; different participants were observed in each task. The first task involved pattern recognition with participants having to locate a target on a screen. The second was a cognitive task in which the participants had to locate a target on a screen. The third task was a driving simulation video game. In each case the response variable was the number of errors committed by a participant. Participants were further divided into three smoking groups: active smokers (AS) composed of people who actively smoked during or just before the task, delayed smokers (DS) composed of regular smokers who had not smoked for 3 hours before the task and non-smokers (NS). The resulting data, taken from Howell (2002), are shown in Table 5.3.

Table 5.2 Survival Times (Days) of Patients with Different Types of Cancer

Stomach	Bronchus	Colon	Ovary	Breast
124	81	248	1234	1235
42	461	377	89	24
25	20	189	201	1581
45	450	1843	356	1166
412	246	180	2970	40
51	166	537	456	727
1112	63	519		3808
46	64	455		791
103	155	406		1804
876	859	365		3460
146	151	942		719
340	166	776		
396	37	372		
	223	163		
	138	101		
	72	20		
	245	283		

5.2 Graphing the Skulls Data

An obvious candidate for the formal analysis of the skulls data in Table 5.1 is *Student's independent samples t-test* (see, for example, Altman, 1991) to assess whether the data give any evidence against the null hypothesis that the population means of the two types of skull are equal. But this test is based on a number of assumptions which should be assessed before applying the test. These assumptions are

■ The measurements are independent of one another.
■ The measurements are normally distributed in each of the two groups.
■ The measurements have the same variance in each group.

Table 5.3 Data from the Effects of Smoking on Performance Study

	Pattern Recognition														
NS	9	8	12	10	7	10	9	11	8	10	8	10	8	11	10
DS	12	7	14	4	8	11	16	17	5	6	9	6	6	7	16
AS	8	8	9	1	9	7	16	19	1	1	22	12	18	8	10
	Cognitive Task														
NS	27	34	19	20	56	35	23	37	4	30	4	42	34	19	49
DS	48	29	34	6	18	63	9	54	28	71	60	54	51	24	49
AS	34	65	55	33	42	54	21	44	61	38	75	61	51	32	47
	Driving Simulation														
NS	15	2	2	14	5	0	16	14	9	17	15	9	3	15	13
DS	7	0	6	0	12	17	1	11	4	4	3	5	16	5	11
AS	3	2	0	0	6	2	0	6	4	1	0	0	6	2	3

Note: NS, non-smoker; DS, delayed smoker; AS, active smoker.

A number of graphical displays might be used to investigate whether or not these assumptions are realistic for the skull data. A straightforward way to graph these data to make it easy to compare the distributions of the head measurements in each group is by a *side-by-side boxplot* and this can be constructed from the following **vbox** plot statement within **proc sgplot**.

```
proc sgplot data=etruscans;
   vbox breadth/group=type nocaps;
   yaxis label='Breadth of skull';
run;
```

We have used the **nocaps** option to suppress the horizontal caps on the ends of the whiskers. The result, shown in Figure 5.1, differentiates the two groups not only by different shades of fill but by different line types, which we do not find appealing. However, to override this would require setting **lineattrs=(pattern=solid)**, **medianattrs=(pattern=solid)** and **whiskerattrs=(pattern=solid)**.

The two boxplots suggest that maximum head breadth tends to be wider for Etruscan skulls than for modern Italian skulls perhaps indicating that Etruscans were not immigrants from Italy but originated from elsewhere. And the boxplots also identify one outlying observation in each group; the exclusion or otherwise of these outliers might need to be carefully considered before any formal analyses of these data. Finally the boxplot for the Italian data indicates that the distribution of head breadths in this group is a little skewed.

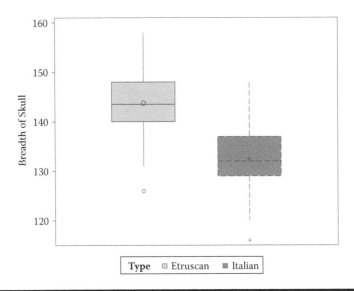

Figure 5.1   Side-by-side boxplots of data on skull breadth in Table 5.1.

We can produce various alternative side-by-side boxplots which may sometimes be more informative than the basic plot demonstrated in Figure 5.1. For example, a vertical arrangement of boxplots is sometimes easier to look at than the horizontal arrangement, particularly when there are several or the labels for them are long. Within **proc sgplot** this is easily done by using **hbox** rather than **vbox**, although here with only two groups to compare the arrangement of boxplots is largely down to individual preference.

```
proc sgplot data=etruscans;
   hbox breadth/group=type nocaps;
   xaxis label='Breadth of skull';
run;
```

The result is shown in Figure 5.2.

We can also compare the distribution of maximum head breadths of Etruscan and Italian skulls by looking at the histograms of each group perhaps enhanced by fitted normal distributions and/or kernel density estimates; **proc sgpanel** can be used to produce such a plot as follows:

```
proc sgpanel data=etruscans;
   panelby type/rows=2 spacing=10 novarname;
   histogram breadth;
   density breadth/type=normal;
   density breadth/type=kernel;
   colaxis label='Breadth of skull';
run;
```

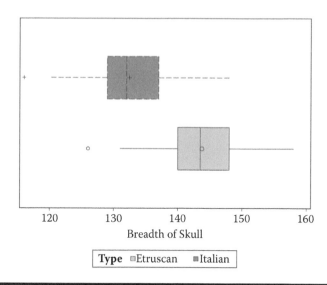

Figure 5.2   Vertical arrangement of boxplots for skull data.

The option rows=2 on the panelby statement ensures that the plots are arranged vertically, rather than side by side, which makes comparison easier. The spacing=10 option adds 10 pixels of space between the panels and novarname uses only the value of type as the title for the panels. The default would include the variable name, e.g. 'Type=Etruscan'. Whereas in proc sgplot, xaxis and yaxis statements are used to control the appearance of the axes, in proc sgpanel colaxis and rowaxis statements are used.

The result is Figure 5.3.

When proc ttest is used to apply an independent samples *t*-test a plot is produced that combines those of Figures 5.2 and 5.3. The following code is used and the plot is shown in Figure 5.4.

```
proc ttest data=etruscans;
   class type;
   var breadth;
run;
```

Figures 5.3 and 5.4 suggest that the head breadth measurements can safely be assumed to have a normal distribution in each group and also indicate that the variation of the observations about the mean is pretty similar in each group. As normality and equal variance are the assumptions made by the two-sample Student's *t*-test, it appears that this test can safely be applied to these data; the results are given by the edited SAS output shown in Table 5.4. (For details of how to use SAS to produce these results, see Der and Everitt, 2013.) The 95% confidence interval

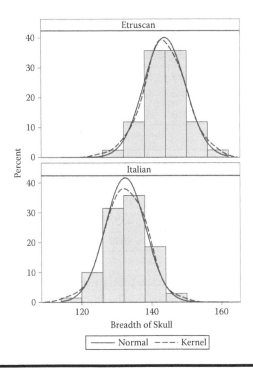

Figure 5.3 Histograms of data on skull measurements in Table 5.1 showing both a fitted normal distribution and a kernel estimate.

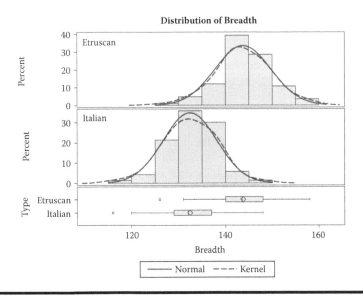

Figure 5.4 Summary panel of skull breadth measurements from proc ttest.

Table 5.4 Results of Student's *t*-Test Applied to the Skull Breadth Data

Type	N	Mean	Std Dev	Std Err	Minimum	Maximum
Etruscan	84	143.8	5.9705	0.6514	126.0	158.0
Italian	70	132.4	5.7499	0.6873	116.0	148.0
Diff (1-2)		11.3310	5.8714	0.9502		

Type	Method	Mean	95% CL Mean		Std Dev	95% CL Std Dev	
Etruscan		143.8	142.5	145.1	5.9705	5.1841	7.0403
Italian		132.4	131.1	133.8	5.7499	4.9301	6.8994
Diff (1-2)	Pooled	11.3310	9.4537	13.2083	5.8714	5.2790	6.6147
Diff (1-2)	Satterthwaite	11.3310	9.4598	13.2021			

for the difference in population means (using the pooled variance estimate which our graphics have justified) is (9.45, 13.21); the evidence from the data is that the mean head breadth of Etruscan skulls is between about 9.5 and 13 mm wider that the Italian skulls.

5.3 Graphing Cancer Survival Times

A possible formal approach to assessing whether or not there is any evidence of a difference in the survival times for the different types of cancer is a *one-way analysis of variance* (see Altman, 1991). This is essentially the analogue of Student's *t*-test when there are more than two groups to compare and is used to test the null hypothesis that the different types of cancer all have the same mean survival time. The assumptions on which the one-way analysis of variance is based are

■ The measurements in each group are normally distributed.
■ The measurements in each group have the same variance.

To investigate these assumptions we shall again begin by looking at a side-by-side boxplot of the data, which can be produced using the following code:

```
data cancers;
  infile 'c: \ hosgus \ data \ patient.dat' 1rec1=50 pad;
  do type=1 to 5;
   input survival 8. @ ;
   if survival ~=. then output;
  end;
run;
```

```
proc format;
   value organ 1='Stomach' 2='Bronchus' 3='Colon' 4='Ovary' 5='Breast';
run;

proc sort data=cancers; by type; run;
proc boxplot data=cancers;
   plot survival*type/boxstyle=schematic noserifs;
   format type organ.;
run;
```

In this example, we use **proc boxplot** which has a wider range of options for boxplots. The syntax is different; the **plot** statement is in the form y* group and the **boxstyle=schematic** option gives a similar appearance to **proc sgplot** but without different attributes for each group. The **noserifs** option is equivalent to the **nocaps** option used with **proc sgplot**.

The resulting plot is given in Figure 5.5.

Figure 5.5 clearly shows that the distribution of survival times for some types of cancer, for example, cancer of the ovaries, is quite skewed. And the figure also indicates the presence of a number of possible outliers in the different categories. The boxplots in Figure 5.5 would suggest that careful thought is needed before applying a one-way analysis of variance directly to these data because the assumptions of normality and homogeneity of variance do not seem to be reasonable here. A simple remedy might be a transformation of the data; here a log

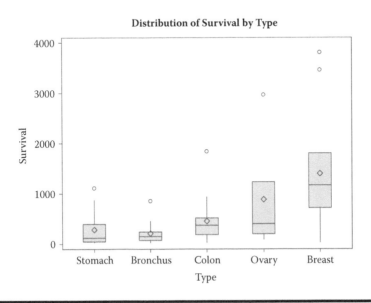

Figure 5.5 Boxplots of survival times (days) for patients with different types of cancer.

transformation suggests itself and we can construct the equivalent diagram to Figure 5.5 but now using the logged survival times from the following SAS code:

```
data cancers;
   set cancers;
   log_survival=log(survival);
run;

proc boxplot data=cancers;
   plot log_survival*type/boxstyle=schematic noserifs;
   format type organ.;
run;
```

The result is shown in Figure 5.6. The distributions of survival time look perhaps a little more 'normal' than those for the original data although not entirely satisfactory and there remain some outliers that are potentially troublesome.

One more variant of the side-by-side box plot that we can illustrate on these data is that which varies box width according to sample size, which can be obtained using the **boxwidthscale** option in **proc boxplot**, as follows:

```
proc boxplot data=cancers;
   plot log_survival*type/boxstyle=schematic boxwidthscale=1 noserifs;
   format type organ.;
run;
```

The result is shown in Figure 5.7; there seems little advantage in this plot over that given in Figure 5.6.

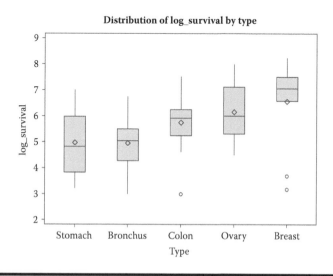

Figure 5.6   Boxplots for the log-transformed survival data.

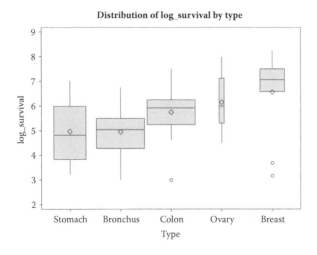

Figure 5.7   Boxplots for survival times data with box width proportional to sample size in the corresponding cancer type.

We will now apply the usual one-way analysis of variance model using SAS (see Der and Everitt, 2013) to both the raw data and the log-transformed data to give the results shown in Table 5.5.

Before we comment on these numerical results we will examine what are known as the *residuals* from both the untransformed and the log-transformed data. The residuals are simply the differences between the observed response value and that predicted from the fitted model. Residuals estimate the error terms in the one-way analysis of variance model and should have a normal distribution. We can use a quantile–quantile (Q–Q) plot for the residuals from the model fitted to the untransformed data and the residuals from the same model fitted to the log-transformed data to assess the normality assumption. These plots are produced as part of a panel of diagnostic plots by **proc glm** and other regression procedures. The components of the diagnostic panel will be described in detail in the next chapter. Here we will unpack the panel into separate plots as follows:

```
proc glm data=cancers plots=diagnostics(unpack);
    class type;
    model survival=type;
run;
proc glm data=cancers plots=diagnostics(unpack);
    class type;
    model log_survival=type;
run;
```

A range of plots is produced but here we give only those in which we are interested, namely, Figures 5.8 and 5.9. The residuals from the log-transformed data are

Table 5.5   One-Way Analysis of Variance Results for Untransformed and Log-Transformed Survival Data

| Untransformed | | | | | |
| | | Sum of | | | |
Source	DF	Squares	Mean Square	F Value	Pr > F
Model	4	11535760.52	2883940.13	6.43	0.0002
Error	59	26448144.48	448273.64		
Corrected Total	63	37983905.00			

| Log-Transformed Data | | | | | |
| | | Sum of | | | |
Source	DF	Squares	Mean Square	F Value	Pr > F
Model	4	24.4865569	6.1216392	4.29	0.0041
Error	59	84.2695894	1.4282981		
Corrected Total	63	108.7561463			

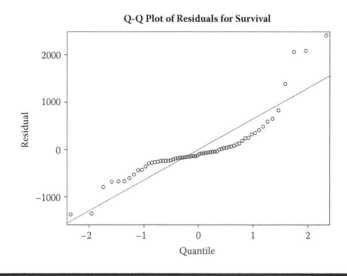

Q-Q Plot of Residuals for Survival

Figure 5.8   Probability plot of residuals from one-way analysis of variance model fitted to the raw survival data.

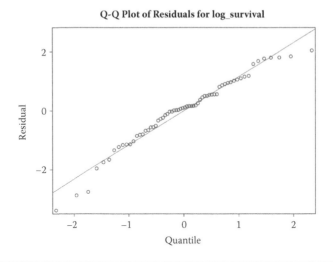

Figure 5.9 **Probability plot of residuals from one-way analysis of variance model fitted to the log-transformed survival data.**

much better behaved than those from the analysis of the raw data. Consequently we will use the results in Table 5.6 on the log-transformed data to draw some conclusions about the data (although here the conclusion from the untransformed data would be much the same). The *F*-test for the null hypothesis of the equality of means has an associated *p*-value of 0.0041 indicating strongly that the null hypothesis is incorrect. Here there is clear evidence of a difference between the population mean log-survival time (and in fact survival time) of the different types of cancer. Figure 5.7 indicates that mean log-survival time is greatest for breast cancer and lowest for stomach and bronchus cancer. But it is not clear from this plot or from the formal test whether *all* cancer types differ in average log-survival time or whether there are only differences between particular cancer types. To uncover which particular types of cancer differ in survival time it is necessary to delve a little further.

Proc glm can be used for a range of analyses that belong to the general linear model, including ANOVA, ANCOVA, and linear regression; it produces a number of plots useful for each type of model. Here we illustrate three plots that can be

Table 5.6 **Analysis of Variance Table for the Smoking Data**

Source	DF	Type I SS	Mean Square	F Value	Pr > F
SmType	2	353.52593	176.76296	1.64	0.1990
Task	2	28620.50370	14310.25185	132.41	<.0001
SmType*Task	4	2727.40741	681.85185	6.31	0.0001

used to examine the differences that may underlie a fitted ANOVA model. Each plot is invoked by the lsmeans statement, as follows:

```
proc glm data=cancers;
  class type;
  model log_survival=type;
  lsmeans type/pdiff=control;
  lsmeans type/pdiff=anom;
  lsmeans type/pdiff=all;
    format type organ.;
run;
```

A boxplot is produced by default and this is shown in Figure 5.10. This is similar to plots shown earlier produced by **proc sgplot** or **proc boxplot**, except that it also contains an inset with the value of the *F* statistic and its associated *p*-value. The first lsmeans statement compares a control group with the remaining groups; the resulting plot is shown in Figure 5.11. Since formatted values of **type** are used the control group is the first level alphabetically. Had we wanted another control group, say **Colon**, the option on the lsmeans statement would be amended to pdiff=control('Colon'). There are also versions designed for one-tailed hypotheses: pdiff=controll where non-controls are expected to be lower than controls and pdiff=controlu where they are expected to be higher. The choice of control group and type of comparison to be made will typically have been chosen when the study was designed.

The second plot compares all cancer types with the overall average and the result is shown in Figure 5.12. The third compares all pairs of cancer types and is shown in Figure 5.13.

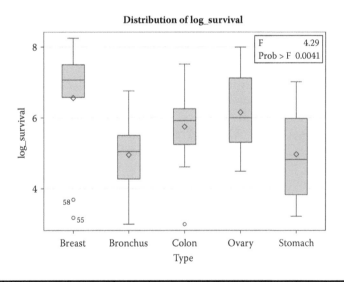

Figure 5.10   Boxplots of log-survival times for cancer data.

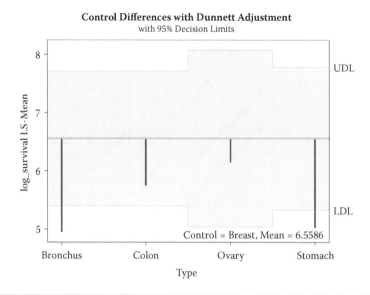

Figure 5.11 Control plot for differences in log survival times for cancer data.

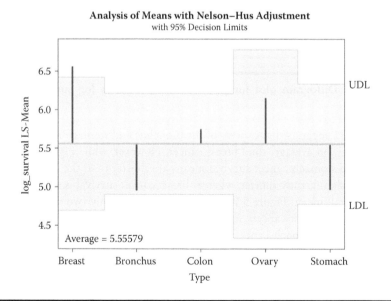

Figure 5.12 Analysis of means plot for log survival times in cancer data.

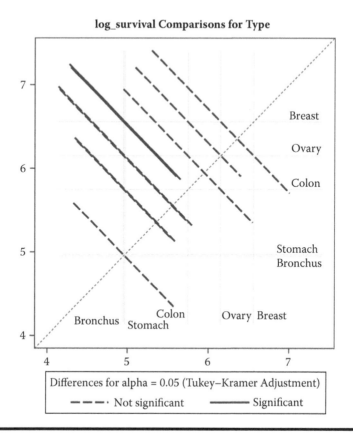

Figure 5.13 Diffogram plot for pairwise differences in log survival times in cancer data.

Figure 5.11 suggests that cancers of the bronchus and stomach result in shorter survival times, on average, than breast cancer. However, while Figure 5.12 shows bronchus and stomach cancer survival means are shorter than the overall average, they are not significantly shorter, whereas breast cancer survival is typically longer than the overall mean. Figure 5.13 shows that there are only two significant pairwise differences: breast cancers survival is significantly longer than that for bronchial cancer and stomach cancer. This plot, known as a mean–mean scatterplot but referred to in SAS as a *diffogram*, needs some explanation. The means of each group are arranged along both the *x* and *y* axes. Each of the 45 degree lines corresponds to a pairwise comparison, namely, the pair whose means intersect at the middle of the line. If a line crosses the dotted diagonal line the pairwise difference is non-significant; significant and non-significant differences are also indicated by different line attributes (pattern and/or colour). In this case, the mean (log) survival for bronchus and stomach cancers is very similar, so their lines are very close.

5.4 Effects of Smoking on Performance

The data in Table 5.3 arise from a 3 × 3 factorial experiment, which would be analysed formally by applying a two-way analysis of variance. The data can be read in and an initial graphic obtained by using the following code:

```
data smoking;
   infile cards stopover;
   input SmType $ @@;
   do i=1 to 15;
      input score @@;
   Task='Pattern';
   if _n_>3 then Task='Cognition';
   if _n_>6 then Task='Driving';
   output;
   end;
   input;
cards;
NS 9 8 12 10 7 10 9 11 8 10 8 10 8 11 10
DS 12 7 14 4 8 11 16 17 5 6 9 6 6 7 16
AS 8 8 9 1 9 7 16 19 1 1 22 12 18 8 10
NS 27 34 19 20 56 35 23 37 4 30 4 42 34 19 49
DS 48 29 34 6 18 63 9 54 28 71 60 54 51 24 49
AS 34 65 55 33 42 54 21 44 61 38 75 61 51 32 47
NS 15 2 2 14 5 0 16 14 9 17 15 9 3 15 13
DS 7 0 6 0 12 17 1 11 4 4 3 5 16 5 11
AS 3 2 0 0 6 2 0 6 4 1 0 0 6 2 3
;

proc sgpanel data=smoking;
   panelby smtype task / layout=lattice uniscale=column;
   vbox score;
run;
```

The resulting plot is shown in Figure 5.14. The large difference in number of errors made in the three tasks is of no importance here (the three tasks could, for example, have been scores in darts, snooker and tiddly-winks); what we are trying to do is assess the effect of smoking on the errors in each task. For the cognition task it appears that the numbers of errors decrease from the active smoking group to the non-smokers, but for the driving task the reverse seems to happen although the differences are small. For the pattern recognition task there seems little difference in the average number of errors in the three groups.

We can now apply two-way analysis of variance to the data (for details of how to do this using SAS, see Der and Everitt, 2013) including the options for the unpacked diagnostic plots and an interaction plot with confidence limits

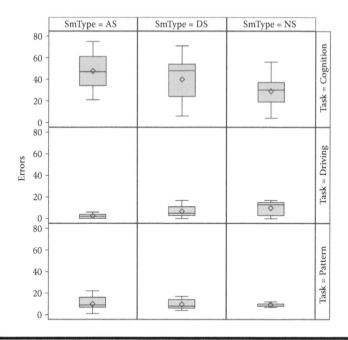

Figure 5.14   Within-cell boxplots for the effect of smoking on performance data.

(essentially a plot of the mean number of errors on each task for the three different types of smoking behaviour).

```
proc glm data=smoking plots(unpack)=(diagnostics intplot(clm));
  class smtype task;
  model errors=smtype|task/solution;
run;
```

The analysis of variance table is shown in Table 5.6, the Q–Q plot in Figure 5.15 and the interaction plot in Figure 5.16.

As a check of the assumption of normality of the error count we can look at a plot of the residuals from the fitted model, shown in Figure 5.15. These are perhaps not as satisfactory as we would wish but we will ignore the possible problem here and carry on to interpret the results.

The highly significant results for the tasks can be ignored for the reasons given earlier. The overall difference in number of errors is not significant but this, of course, conceals the fact that there is a highly significant interaction effect which provides evidence that the effect of smoking on the number of errors made differs among the three tasks. This interaction effect is shown graphically in Figure 5.16. Clearly smoking had an adverse effect on the number of errors made in the cognitive task but little effect on the number of errors made in the other two tasks.

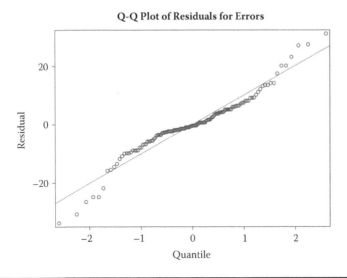

Figure 5.15 Normal probability plot of residuals from the two-way analysis of variance model for the smoking data.

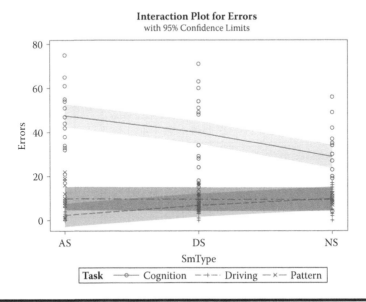

Figure 5.16 Interaction plot for the smoking data.

5.5 Summary

A continuous variable grouped into categories by one or more nominal scale variables can usefully be displayed in a variety of ways and such plots can give a useful initial view of the data and can also help in the assessment of the assumptions of tests like the *t*-test and procedures like the analysis of variance that may be applied to the data. Graphs can also be helpful in a more detailed investigation of group differences in an analysis of variance.

Exercises

5.1. Shortly after metric units of length were officially introduced in Australia in the 1970s, each of a group of 44 students was asked to guess, to the nearest metre, the width of the lecture hall in which they were sitting. Another group of 69 students in the same room was asked to guess the width in feet, to the nearest foot. The data are given in Table 5.7. The main question of interest is whether estimation in feet and in metres gives different results. Investigate this question using suitable graphical displays, followed by what you consider the appropriate formal test.

Table 5.7   Room Width Estimates

Guesses in Metres											
8	9	10	10	10	10	10	10	11	11	11	11
12	12	13	13	13	14	14	14	15	15	15	15
15	15	15	15	16	16	16	17	17	17	17	18
18	20	22	25	27	35	38	40				
Guesses in Feet											
24	25	27	30	30	30	30	30	30	32	32	33
34	34	34	35	35	36	36	36	37	37	40	40
40	40	40	40	40	40	40	41	41	42	42	42
42	43	43	44	44	44	45	45	45	45	45	45
46	46	47	48	48	50	50	50	51	54	54	54
55	55	60	60	63	70	75	80	94			

5.2. After the students in Exercise 5.1 had made estimates of the width of the lecture hall, the room was accurately measured and found to be 13.1 metres (43.0 feet). Add this additional information to the graphics you produced in the first exercise. Does the information help to determine which of the two types of estimates was more precise?

5.3. The data in Table 5.8 give the percentage of household with various facilities and equipment in four areas of East Jerusalem in 1967. Construct some graphics that may help in assessing similarities and differences among the different areas and to see how patterns of access to facilities differ among them.

Table 5.8   Percentages of Households with Various Facilities and Equipment in Four Areas of East Jerusalem in 1967

	Christian	Armenian	Jewish	Moslem
Toilet	98.2	97.2	97.3	96.9
Kitchen	78.8	81.0	65.6	73.3
Bath	14.4	17.6	6.0	9.6
Electricity	86.2	82.1	54.5	74.7
Water	32.9	30.3	21.1	26.9
Radio	73.0	70.4	53.0	60.5
TV set	4.6	6.0	1.5	3.4
Refrigerator	29.2	26.3	4.3	10.5

5.4. The data in Table 5.9 give the supine systolic and diastolic blood pressures (mm Hg) for 35 patients with moderate essential hypertension, immediately before and two hours after taking a drug, captopril (the data are from MacGregor et al., 1979). The interest is in investigating the response to the drug treatment. What graphics do you think would be helpful in throwing light on this question? After constructing what you think are the relevant plots, use them to direct you to a more formal test of the effect of the drug.

5.5. The silver content (%Ag) of a number of Byzantine coins discovered in Cyprus was determined. Nine of the coins came from the first coinage of the reign of King Manuel I, Comenus (1143–1180); there were seven from the second coinage minted several years later and four from the third coinage (later still); another seven were from a fourth coinage. The question arose of whether there were significant differences in the silver content of coins minted early and late in Manuel's reign. The data are given in Table 5.10. Use suitable graphics to suggest an appropriate method of analysis to assess possible difference in the silver content of the various groups of coins, carry out the formal analysis and then graphically assess the assumptions of the model which you have used.

Table 5.9   Captopril and Blood Pressure

	Systolic		Diastolic	
Patient ID	Before	After	Before	After
1	210	201	130	125
2	169	165	122	121
3	187	166	124	121
4	160	157	104	106
5	167	147	112	101
6	176	145	101	85
7	185	168	121	98
8	206	180	124	105
9	173	147	115	103
10	146	136	102	98
11	174	151	98	90
12	201	168	119	98
13	198	179	106	110
14	148	129	107	103
15	154	131	100	82

Table 5.10   Silver Content of Byzantine Coins

First	Second	Third	Fourth
5.9	6.9	4.9	5.3
6.8	9.0	5.5	5.6
6.4	6.6	4.6	5.5
7.0	8.1	4.5	5.1
6.6	9.3		6.2
7.7	9.2		5.8
7.2	8.6		5.8
6.9			
6.2			

5.6. The set of data shown in Table 5.11 gives the heights in inches of male singers in the New York Choral Society grouped according to voice parts with the vocal range decreasing in pitch going from Tenor 1 to Bass 2. Interest here lies in whether there is any evidence that there is a difference in height for singers of different pitch. Investigate the data graphically before conducting a formal test of the hypothesis that different types of singers have the same population mean; if required use some graphical displays to investigate which voice types differ in height.

Table 5.11 Heights of Male Singers by Vocal Range

Tenor 1	Tenor 2	Bass 1	Bass 2
69	68	72	72
72	73	70	75
71	69	72	67
66	71	69	75
76	69	73	74
74	76	71	72
71	71	72	72
66	69	68	74
68	71	68	72
67	66	71	72
70	69	66	74
65	71	68	70
72	71	71	66
70	71	73	68
68	69	73	75
64	70	70	68
73	69	68	70
66	68	70	72
68	70	75	67

Continued

Table 5.11 (*Continued*) Heights of Male Singers by Vocal Range

Tenor 1	Tenor 2	Bass 1	Bass 2
67	68	68	70
64	69	71	70
		70	69
		74	72
		70	71
		75	74
		75	75
		69	
		72	
		71	
		70	
		71	
		68	
		70	
		75	
		72	

5.7. Reanalyse the smoking data after taking a square root transformation of the error count.

Chapter 6

Linear Regression, the Scatterplot and Beyond: Galaxies, Anaerobic Threshold, Birds on Islands, Birth and Death Rates, US Birth Rates during and after World War II and Air Pollution in US Cities

6.1 Introduction

In this chapter we will look at the following six datasets:

Galaxies—Wood (2006) gives the relative velocity and the distance of 24 galaxies, according to measurements made using the Hubble Space Telescope (see Table 6.1). Velocities are assessed by measuring the Doppler red shift in the spectrum of light observed from the galaxies concerned, although some correction

Table 6.1 Distance and Velocity for 24 Galaxies

Observation	Galaxy	Velocity (kms)	Distance (megaparsec)
1	NGC0300	133	2.00
2	NGC0925	664	9.16
3	NGC1326A	1794	16.14
4	NGC1365	1594	17.95
5	NGC1425	1473	21.88
6	NGC2403	278	3.22
7	NGC2541	714	11.22
8	NGC2090	882	11.75
9	NGC3031	80	3.63
10	NGC3198	772	13.80
11	NGC3351	642	10.00
12	NGC3368	768	10.52
13	NGC3621	609	6.64
14	NGC4321	1433	15.21
15	NGC4414	619	17.70
16	NGC4496A	1424	14.86
17	NGC4548	1384	16.22
18	NGC4535	1444	15.78
19	NGC4536	1423	14.93
20	NGC4639	1403	21.98
21	NGC4725	1103	12.36
22	IC4182	318	4.49
23	NGC5253	232	3.15
24	NGC7331	999	14.72

for 'local' velocity components is required. Distances are measured using the known relationship between the period of Cepheid variable stars and their luminosity. How can these data be used to estimate the age of the universe?

Anaerobic threshold—Bennett (1988) describes data collected during an experiment in kinesiology. A subject performed a standard exercise task at a gradually increasing level and recorded oxygen uptake and expired ventilation. The resulting data are shown in Table 6.2. The question is how these variables are related.

Birds on islands—Vuilleumier (1970) in an investigation of the numbers of bird species in isolated islands of paramo vegetation in the northern Andes collected the data shown in Table 6.3. The aim here is to investigate how the number of species is related to the other two variables.

Table 6.2 Data on Oxygen Uptake and Expired Ventilation

Oxygen Uptake	Expired Ventilation	Oxygen Uptake	Expired Ventilation
574	21.9	2577	46.3
592	18.6	2766	55.8
664	18.6	2812	54.5
667	19.1	2893	63.5
718	19.2	2957	60.3
770	16.9	3052	64.8
927	18.3	3151	69.2
947	17.2	3161	74.7
1020	19.0	3266	72.9
1096	19.0	3386	80.4
1277	18.6	3452	83.0
1323	22.8	3521	86.0
1330	24.6	3543	88.9
1599	24.9	3676	96.8
1639	29.2	3741	89.1
1787	32.0	3844	100.9
1790	27.9	3878	103.0
1794	31.0	4002	113.4

Continued

Table 6.2 (Continued)   Data on Oxygen Uptake and Expired Ventilation

Oxygen Uptake	Expired Ventilation	Oxygen Uptake	Expired Ventilation
1874	30.7	4114	111.4
2049	35.4	4152	119.9
2132	36.1	4252	127.2
2160	39.1	4290	126.4
2292	42.6	4331	135.5
2312	39.9	4332	138.9
2475	46.2	4390	143.7
2489	50.9	4393	144.8
2490	46.5		

Table 6.3   Birds in Paramo Vegetation

Island	N	Area	Elevation
Chiles	36	0.33	1.26
Las Papas-Coconuco	30	0.50	1.17
Sumapaz	37	2.03	1.06
Tolima-Quindio	35	0.99	1.90
Paramillo	11	0.03	0.46
Cocuy	21	2.17	2.00
Pamplona	11	0.22	0.70
Cachira	13	0.14	0.74
Tama	17	0.05	0.61
Batallon	13	0.07	0.66
Merida	29	1.80	1.50
Perija	4	0.17	0.75
Santa Marta	18	0.61	2.28
Cende	15	0.07	0.55

Birth and death rates—Table 6.4 gives the birth and death rates for 69 countries. Here we might perhaps like to 'explore' the data for any evidence of distinct groups of countries.

Birth rates—Cook and Weisberg (1982) give the monthly US births per thousand population for the years 1940 to 1948; the data are shown in Table 6.5. Here we will demonstrate how changing various features of the scatterplot can make the data more transparent.

Air pollution in US cities—Data were collected by Sokal and Rohlf (1981) from several US government publications on the annual mean concentration of sulphur dioxide, in micrograms per cubic metre, in the air in 41 US cities; this measurement reflects the air pollution in a city. The values of six other variables

Table 6.4 Birth and Death Rates for 69 Countries

	Birth	*Death*
Alg	36.4	14.6
Con	37.3	8.0
Egy	42.1	15.3
Gha	55.8	25.6
Ict	56.1	33.1
Mag	41.8	15.8
Mor	46.1	18.7
Tun	41.7	10.1
Cam	41.4	19.7
Cey	35.8	8.5
Chi	34.0	11.0
Tai	36.3	6.1
Hkg	32.1	5.5
Ind	20.9	8.8
Ids	27.7	10.2
Irq	20.5	3.9
Isr	25.0	6.2

Continued

Table 6.4 (*Continued*) Birth and Death Rates for 69 Countries

	Birth	Death
Jap	17.3	7.0
Jor	46.3	6.4
Kor	14.8	5.7
Mal	33.5	6.4
Mog	39.2	11.2
Phl	28.4	7.1
Syr	26.2	4.3
Tha	34.8	7.9
Vit	23.4	5.1
Can	24.8	7.8
Cra	49.9	8.5
Dmr	33.0	8.4
Gut	47.7	17.3
Hon	46.6	9.7
Mex	46.1	10.5
Nic	42.9	7.1
Pan	40.1	8.0
Usa	21.7	9.6
Arg	21.8	8.1
Bol	17.4	5.8
Bra	45.0	13.5
Chl	33.6	11.8
Clo	44.0	11.7
Ecu	44.2	13.5
Per	27.7	8.2
Urg	22.5	7.8

Table 6.4 (*Continued*)   Birth and Death Rates for 69 Countries

	Birth	*Death*
Ven	42.8	6.7
Aus	18.8	12.8
Bel	17.1	12.7
Brt	18.2	12.2
Bul	16.4	8.2
Cze	16.9	9.5
Dem	17.6	19.8
Fin	18.1	9.2
Fra	18.2	11.7
Gmy	18.0	12.5
Gre	17.4	7.8
Hun	13.1	9.9
Irl	22.3	11.9
Ity	19.0	10.2
Net	20.9	8.0
Now	17.5	10.0
Pol	19.0	7.5
Pog	23.5	10.8
Rom	15.7	8.3
Spa	21.5	9.1
Swe	14.8	10.1
Swz	18.9	9.6
Rus	21.2	7.2
Yug	21.4	8.9
Ast	21.6	8.7
Nzl	25.5	8.8

Table 6.5 US Monthly Birth Rates between 1940 and 1948

1890	1957	1925	1885	1896	1934	2036	2069	2060
1922	1854	1852	1952	2011	2015	1971	1883	2070
2221	2173	2105	1962	1951	1975	2092	2148	2114
2013	1986	2088	2218	2312	2462	2455	2357	2309
2398	2400	2331	2222	2156	2256	2352	2371	2356
2211	2108	2069	2123	2147	2050	1977	1993	2134
2275	2262	2194	2109	2114	2086	2089	2097	2036
1957	1953	2039	2116	2134	2142	2023	1972	1942
1931	1980	1977	1972	2017	2161	2468	2691	2890
2913	2940	2870	2911	2832	2774	2568	2574	2641
2691	2698	2701	2596	2503	2424			

Note: Read along rows for temporal sequence.

were also collected, two of which relate to human ecology (i.e. population size in thousands and number of manufacturing enterprises employing more than 20 workers) and four to climate (i.e. average annual temperature [F], average annual wind speed in miles per hour, average annual precipitation in inches and average number of days with precipitation per year). The data are given in Table 6.6. In this example there are seven variables one of which, sulphur dioxide concentration, is a 'response', and the other six are 'explanatory' variables (often wrongly labelled 'independent' variables). Interest here lies in determining how the ecology and climate variables affect sulphur dioxide concentration.

Table 6.6 Data on Air Pollution in 41 US Cities

City	SO_2	Temperature	Factories	Population	Wind Speed	Rain	Rainy Days
Phoenix	10	70.3	213	582	6.0	7.05	36
Little Rock	13	61.0	91	132	8.2	48.52	100
San Francisco	12	56.7	453	716	8.7	20.66	67
Denver	17	51.9	454	515	9.0	12.95	86
Hartford	56	49.1	412	158	9.0	43.37	127

Table 6.6 (*Continued*) Data on Air Pollution in 41 US Cities

City	SO$_2$	Temperature	Factories	Population	Wind Speed	Rain	Rainy Days
Wilmington	36	54.0	80	80	9.0	40.25	114
Washington	29	57.3	434	757	9.3	38.89	111
Jacksonville	14	68.4	136	529	8.8	54.47	116
Miami	10	75.5	207	335	9.0	59.80	128
Atlanta	24	61.5	368	497	9.1	48.34	115
Chicago	110	50.6	3344	3369	10.4	34.44	122
Indianapolis	28	52.3	361	746	9.7	38.74	121
Des Moines	17	49.0	104	201	11.2	30.85	103
Wichita	8	56.6	125	277	12.7	30.58	82
Louisville	30	55.6	291	593	8.3	43.11	123
New Orleans	9	68.3	204	361	8.4	56.77	113
Baltimore	47	55.0	625	905	9.6	41.31	111
Detroit	35	49.9	1064	1513	10.1	30.96	129
Minneapolis	29	43.5	699	744	10.6	25.94	137
Kansas City	14	54.5	381	507	10.0	37.00	99
St. Louis	56	55.9	775	622	9.5	35.89	105
Omaha	14	51.5	181	347	10.9	30.18	98
Albuquerque	11	56.8	46	244	8.9	7.77	58
Albany	46	47.6	44	116	8.8	33.36	135
Buffalo	11	47.1	391	463	12.4	36.11	166
Cincinnati	23	54.0	462	453	7.1	39.04	132
Cleveland	65	49.7	1007	751	10.9	34.99	155
Columbus	26	51.5	266	540	8.6	37.01	134
Philadelphia	69	54.6	1692	1950	9.6	39.93	115
Pittsburgh	61	50.4	347	520	9.4	36.22	147

Continued

Table 6.6 (*Continued*)   Data on Air Pollution in 41 US Cities

City	SO$_2$	Temperature	Factories	Population	Wind Speed	Rain	Rainy Days
Providence	94	50.0	343	179	10.6	42.75	125
Memphis	10	61.6	337	624	9.2	49.10	105
Nashville	18	59.4	275	448	7.9	46.00	119
Dallas	9	66.2	641	844	10.9	35.94	78
Houston	10	68.9	721	1233	10.8	48.19	103
Salt Lake City	28	51.0	137	176	8.7	15.17	89
Norfolk	31	59.3	96	308	10.6	44.68	116
Richmond	26	57.8	197	299	7.6	42.59	115
Seattle	29	51.1	379	531	9.4	38.79	164
Charleston	31	55.2	35	71	6.5	40.75	148
Milwaukee	16	45.7	569	717	11.8	29.07	123

6.2 The Scatterplot

The basic graphic for two continuous variables is the simple *xy* scatterplot, which has been in use since at least the 18th century and is still widely used in initial explorations of bivariate data. According to Tufte (1983):

> The relational graphic—in its barest form the scatterplot and its variants— is the greatest of all graphical designs. It links at least two variables encouraging and even imploring the viewer to assess the possible causal relationship between the plotted variables. It confronts causal theories that *x* causes *y* with empirical evidence as to the actual relationship between *x* and *y*.

The ubiquitous scatterplot is fundamental in the analysis of bivariate data particularly when enhanced in various ways; it should always be produced and studied before undertaking any formal analyses of such data. In many cases the scatterplot will suggest the most appropriate way or ways to actually analyse the data. The capacity of a scatterplot to convey information about bivariate data accounts for its popularity. In this chapter we shall illustrate the use of the scatterplot and demonstrate how it can be enhanced in a number of ways to make it even more useful.

6.3 Galaxies

We can produce the basic scatterplot of velocity against distance for the data in Table 6.1 using the following code:

```
proc sgplot data=universe;
  scatter y=velocity x=distance;
run;
```

The result is shown in Figure 6.1. The diagram shows a clear, strong relationship between velocity and distance with larger velocities tending to be associated with greater distances. But Cleveland (1993) makes the point that there are two components to visualizing the structure of statistical data: *graphing* and *fitting*. Graphs are needed, of course, because visualization implies a process in which information is encoded on visual displays. Fitting mathematical functions to data is also often needed, otherwise important aspects of data may remain undiscovered. And scatterplots are almost always more interesting if some estimated fit of the relationship between the two variables is shown.

The basic scatterplot of the galaxy data can be enhanced so as to display the likely form of the velocity–distance relationship by showing the simple linear regression fit; in this case, however, we have to remember that the nature of the data requires a model *without* an intercept because if distance is zero, so is relative speed. So the model fitted to the data is simply

$$\text{Velocity} = \beta \text{Distance} + \text{Error} \tag{6.1}$$

(This is essentially what astronomers call *Hubble's law*.) The error terms are assumed to be normally distributed with constant variance, i.e. a variance that

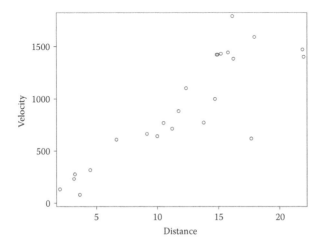

Figure 6.1   Scatterplot of the galaxy data in Table 6.1.

Table 6.7 Results of Fitting a Linear Regression Model with Zero Intercept to the Galaxy Data

Parameter Estimates					
Variable	DF	Parameter Estimate	Standard Error	t Value	Pr > |t|
Distance	1	76.58117	3.96479	19.32	<.0001

does not depend on distance. The model can easily be fitted using SAS (see Der and Everitt, 2009, Chapter 6, for details) and gives the results shown in Table 6.7.

We can now plot the fitted line onto the scatterplot along with the upper and lower bounds of the 95% confidence interval for the line. To do this we use proc reg (we could equally have used proc glm).

```
proc reg data=universe plots=fitplot;
  model velocity=distance/noint;
run;
```

Although scatterplots with regression lines fitted can be produced using proc sgplot, as in later examples in this chapter, it would not be possible to constrain the intercept to pass through the origin as is done here with the noint option on the model statement. With ODS enabled, the fitplot is produced by default, so the plots=fitplot option is not strictly necessary here.

The result is shown in Figure 6.2. The square of the correlation coefficient between the two variables (R-square) indicates that 94% of the variance in velocity is accounted for by distance; the linear regression line appears to fit the data very well.

A regression analysis should not end without an attempt to check assumptions such as those of constant variance and normality of the error terms. Violation of these assumptions may invalidate conclusions based on the regression analysis. The estimated *residuals*, i.e. the differences between an observed response (y_i) and the value predicted by the fitted model $(\hat{y}_i)$, $r_i = y_i - \hat{y}_i$, play an essential role in diagnosing a fitted model. But because the basic residuals do not have the same variance (the precision of $\hat{y}_i$ depends upon x_i), they are sometimes standardized before use. There are two possibilities: the *standardized residual* and the *studentized residual*. These are defined as follows:

$$\text{Standardized residual: } r_{\text{sta}} = \frac{y_i - \hat{y}_i}{s\sqrt{1 - h_i}} \tag{6.2}$$

$$\text{Studentized residual: } r_{\text{stu}} = \frac{y_i - \hat{y}_i}{s_{(-i)}\sqrt{1 - h_i}} \tag{6.3}$$

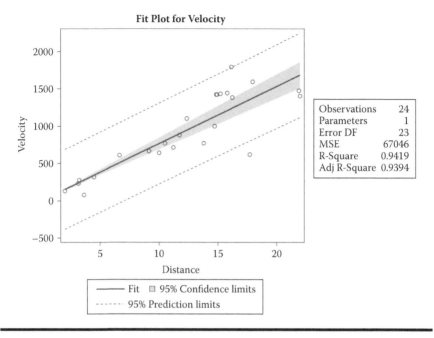

Figure 6.2 Scatterplot of galaxy data showing fitted zero intercept linear regression.

where h_i is the ith diagonal element of the so-called 'hat matrix' (see Cook and Weisberg, 1982) and $s^2_{(-i)}$ is the estimated residual variance from fitting the model after the exclusion of the ith observation. The following diagnostic plots using one or the other of the residual terms are generally helpful when assessing model assumptions:

■ Residuals versus fitted values – If the fitted model is appropriate, the plotted points should lie in an approximately horizontal band across the plot. Departures from this appearance may indicate that the functional form of the assumed model is incorrect or, alternatively, that there is non-constant variance.
■ Residuals versus explanatory variables – Systematic patterns in these plots can indicate violations of the constant variance assumption or an inappropriate model form.
■ Normal probability plot of the residuals – The plot checks the normal distribution assumptions on which all statistical inferences procedures are based.

Figure 6.3 shows some idealized plots that indicate particular points about models. Figure 6.3a is what is looked for to confirm that the fitted model meets the assumptions of the regression model, Figure 6.3b suggests that the assumption of constant variance is not justified so a transformation of the response variable before fitting might be a sensible option to consider and Figure 6.3c implies that the model requires a quadratic term in the explanatory variables used in the plot.

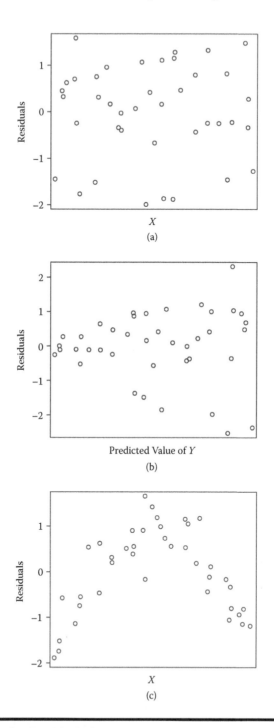

Figure 6.3   Idealized residual plots.

A further diagnostic that is often very useful is an index plot of the Cook's distances for each observation. This statistic is defined as follows:

$$D_k = \frac{1}{(p+1)s^2} \sum_{i=1}^{n} \left[\hat{y}_{i(k)} - \hat{y}_i \right]^2 \qquad (6.4)$$

where $\hat{y}_{i(k)}$ is the fitted value of the ith observation when the kth observation is omitted from the model. The values of D_k assess the impact of the kth observation on the estimated regression coefficients. Values of D_k greater than one are suggestive that the corresponding observation has undue influence on the estimated regression coefficients (again see Cook and Weisberg, 1982).

When ODS graphics are enabled, a panel of useful diagnostic plots is produced by default. Here we have requested the plots explicitly in order to use the label option to identify any influential points or outliers. The variable named on the id statement is used by the label option to identify the points in the plots. The code is

```
proc reg data=universe plots=(diagnostics(label));
  model velocity=distance/noint;
  id id;
run;
```

The plot is shown in Figure 6.4. Although it is generally useful to have a range of plots in a single panel like this, there may be occasions when these small plots do not offer sufficient detail, for example with large datasets or numerous labelled points. In these cases we can request separate plots by including the unpack option; the plot request would then be plots=(diagnostics(label unpack)).

By default a plot of residuals against the predictor, distance, is also produced. This is shown in Figure 6.5.

First let us look at Figure 6.4. The figure contains the following plots:

■ Residuals versus the predicted values
■ Externally studentized residuals (rstudent) versus the predicted values
■ Externally studentized residuals versus the leverage
■ Normal quantile–quantile plot (Q–Q plot) of the residuals
■ Dependent variable values versus the predicted values
■ Cook's D versus observation number
■ Histogram of the residuals
■ 'Residual-Fit' (or RF) plot consisting of side-by-side quantile plots of the centred fit and the residuals

(If you specify the stats=none sub-option, then a boxplot of the residuals is produced.)

The plots for assessing the normality assumption for the error terms in the model are the Q–Q plot of residuals and the histogram of the residuals;

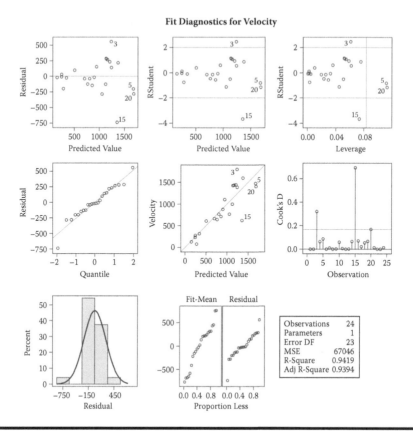

Figure 6.4 **Residual plots given by SAS for galaxy data.**

in Figure 6.4 both of these indicate that the normality assumption is reasonable. For checking the constant variance assumption, plots of residuals (raw or studentized) against the predicted values and residuals against distance (Figure 6.5) can be used. Here these three plots appear to indicate a problem with the assumption; the residuals have increasing variance with increasing predicted values. Next the studentized residual plot can be used to identify outliers in the data, which in this context means observations that have undue *influence* of the estimated regression coefficient; observations whose studentized residuals lie outside the band between the reference lines rstudent = ±2 are deemed outliers. Similarly in the leverage plot observations whose leverage values are greater than the vertical reference leverage = $2p/n$, where p is the number of parameters including the intercept and n is the number of observations used, may also be overly influential in the estimation of the regression coefficient. (See Rawlings et al., 2001, for more details.).

Finally observations whose Cook's D statistic value lies above the horizontal reference line at value $4/n$, where n is the number of observations used, are also

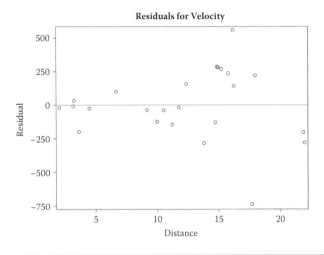

Figure 6.5   Plot of residuals against distance for galaxy data.

deemed to be influential (again see Rawlings et al., 2001). Examination of the relevant plots in Figure 6.4 suggests that galaxies 3, 15 and perhaps 20 could be causing problems in the estimation of the model.

Taken together the plots in Figures 6.4 and 6.5 suggest that it might be sensible to redo the regression analysis of the data after removing galaxies 3, 15 and 20; in addition, investigation of a weighted regression approach might be considered. We leave both of these as exercises for the readers (see Exercise 6.1).

Finally we cannot leave this example without showing how the estimated regression coefficient can be used to estimate the age of the universe. The Hubble constant itself has units of $(km)s^{-1} (Mpc)^{-1}$. A megaparsec is 3.09×10^{19} km, so we need to divide the estimated value of β by this amount in order to obtain Hubble's constant with units of s^{-1}. The approximate of the universe in seconds will then be $1/\beta$. Carrying out the necessary calculations gives an estimated age of 12,794,888,643 years.

6.4 Anaerobic Threshold

Adding a simple linear fit to a scatterplot as in the previous example is not all that can be done to help understand the relationship between two continuous variables. Often a parametric regression model, whilst very useful, does not provide an adequate account of a dataset and frequently a useful alternative (or addition) is to estimate the fit *locally* so that at any point the fit at that point depends only on the observations at that point and some specified neighbouring points. Because such a fit produces an estimate of the *y* variable that is less variable than the original observed value, the result is often called a *smooth* and procedures for producing such fits are called *scatterplot smoothers*. One of the most commonly used scatterplot smoothers is the *loess fit* where we assume that the relationship between the two variables is

$$y_i = g(x_i) + \varepsilon_i \tag{6.5}$$

where g is a 'smooth' function and the ε_i are random variables with zero mean and constant scale. Fixed values $\hat{y}_i$ are used to estimate the y_i at each x_i by fitting polynomials using weighted least squares with large weights for points near to x_i and small otherwise. Two parameters need to be chosen in loess fitting. The first is a smoothing parameter with large values leading to smoother fits and the second is the degree of certain polynomials that are used in the fitting process. For full details, see Cleveland (1979).

So to begin our investigation of the data in Table 6.2 we will construct the scatterplot of the two variables with added linear and loess fits. The code needed is as follows:

```
proc sgplot data=anaerob;
  reg y=exp x=o2in;
  loess y=exp x=o2in/nomarkers;
run;
```

Using the **reg** plot statement gives a scatter plot as well as the regression line, so a separate **scatter** statement is not necessary. Since the **loess** statement also produces a scatterplot the **nomarkers** option has been used to suppress this; otherwise two different markers would have been superimposed on the points.

The resulting plot is shown in Figure 6.6.

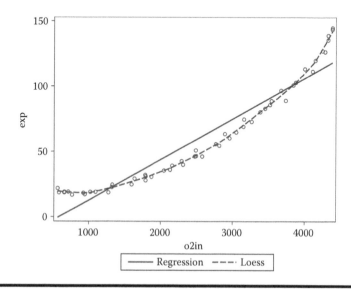

Figure 6.6   Scatterplot of expired ventilation data showing linear and loess fits.

Clearly a linear fit is not appropriate. A more complicated model is needed to represent the data adequately. An obvious choice is to consider a model that in addition to the linear effect of oxygen uptake includes a quadratic term in this variable, i.e. the following model:

$$y_i = \beta_0 + \beta_1 x_i + \beta_2 x_i^2 + \varepsilon_i \qquad (6.6)$$

How this model is fitted using SAS is described in Der and Everitt (2013). Here we simply use **proc sgplot** to display the equivalent fitted values of expired ventilation on the scatterplot of the data along with the loess smooth, as follows:

```
proc sgplot data=anaerob;
  reg y=exp x=o2in/degree=2;
  loess y=exp x=o2in/nomarkers;
run;
```

In addition to its default use for a linear fit, the **reg** statement within **proc sgplot** can be used to display polynomial fits via the **degree=** option; **degree=2** for a quadratic fit, **3** for a cubic fit and so on. The result is shown in Figure 6.7. Clearly the quadratic model fits the data very well.

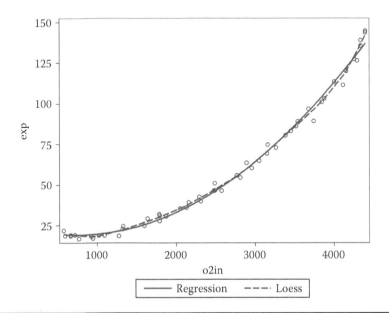

Figure 6.7 Scatterplot of expired ventilation data showing fitted quadratic regression.

6.5 Birds on Islands

For the birds on islands data there are three variables of interest but rather than producing three scatterplots, one for each pair of the variables, we will construct what is known as a *bubble plot*, a plot in which two variables form the basic scatterplot and the values of the third variable are represented by the radii of circles centred at the appropriate place on the scatterplot, i.e. the point defined by the first two variable values of an observation. We will also label each point with the name of the island from which the corresponding observations arose. The plot is obtained using the code:

```
proc sgplot data=paramo;
  bubble y=n x=elevation size=area/datalabel=island;
run;
```

The resulting bubble plot is shown in Figure 6.8. The plot clearly demonstrates that the islands fall into two distinct groups of islands: one group consists of islands that are smaller in area, lower in elevation and have fewer species than the members of the second group. In the group of larger islands, increasing elevation is associated with decreasing number of species; in the other group this relationship is unclear, primarily because the islands all have rather similar elevations. Although there are relatively few islands in each group it might still be informative to show on the scatterplot the linear and locally weighted fits calculated separately from the islands in the two obvious groups identified in the scatterplot.

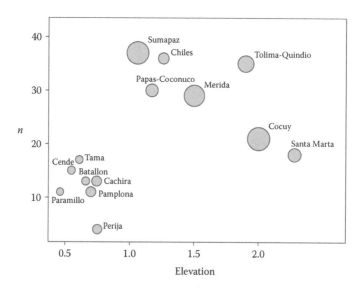

Figure 6.8   Bubble plot of the data on number of bird species on different islands.

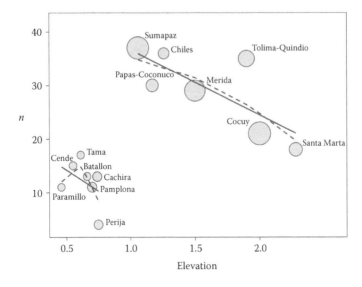

Figure 6.9 Bubble plot of number of bird species on different islands showing linear and locally weighted regression fits in the two groups of islands identified in the original scatterplot.

The code needed is

```
proc sgplot data=paramo noautolegend;
  bubble y=n x=elevation size=area/datalabel=island;
  reg y=n x=elevation/group=group nomarkers lineattrs=(pattern=solid);
  loess y=n x=elevation/group=group nomarkers lineattrs=(pattern=dash);
run;
```

The plot is shown in Figure 6.9. Clearly as elevation increases in both groups of islands, the number of bird species decreases; the difference between the linear fit and the locally weighted fit in the smaller islands is largely the result of the small number of islands in the group.

6.6 Birth and Death Rates

Examination of scatterplots often centres on assessing density patterns such as clusters, gaps or outliers. But humans are not particularly good at visually examining point density, and some type of density estimate added to the scatterplot will frequently be very helpful; this may be particularly helpful in indicating whether the data contain regions of relatively high density separated by regions of relatively low density suggesting perhaps the presence of 'clusters' in the data, which might then be uncovered more formally using some form of *cluster analysis* (see Everitt et al., 2011).

Bivariate density estimation is essentially a simple extension to the univariate method described in Chapter 3. (There is now a vast literature on density estimation, see, for example, Silverman, 1986, and here we give only a very brief summary.)

The univariate kernel density estimator considered in Chapter 3 has a straightforward extension to two dimensions and the bivariate estimator for a set of n sample values, (X_1, Y_1), (X_2, Y_2), ..., (X_n, Y_n) at a point (x, y) is defined as

$$\hat{f}(x, y) = \frac{1}{nh_x h_y} \sum_{i=1}^{n} K\left(\frac{x - X_i}{h_x}, \frac{y - Y_i}{h_y}\right)$$

(6.7)

In this estimator each coordinate direction has its own smoothing parameter, h_x and h_y. An alternative is to scale the data equally for both dimensions and use a single smoothing parameter. In bivariate density estimation the kernel function K is often the standard bivariate normal density given by

$$K(x, y) = \frac{1}{2\pi} \exp\left[-\frac{1}{2}(x^2 + y^2)\right]$$

An alternative is the bivariate Epanechikov kernel given by

$$K(x, y) = \frac{2}{\pi}(1 - x^2 + y^2) \text{ if } x^2 + y^2 < 1$$

$$= 0 \text{ otherwise}$$

We can illustrate enhancing a scatterplot with the estimated bivariate density of the two variables on the birth and death rate data in Table 6.4. The required code is

```
proc kde data=fertility;
  bivar birth death/bwm=0.5 plots=contourscatter;
run;
```

The plot is shown in Figure 6.10 and suggests perhaps two groups of observations, which on examination correspond largely to Western and Third World countries. As mentioned earlier the possible clustering in the data could now be investigated more formally using one of the many methods of cluster analysis available. (For an account of cluster analysis, see Everitt et al., 2011.)

Rather than the contour plot used in Figure 6.10 we could use a *perspective plot* given by the following code:

```
proc kde data=fertility;
  bivar death birth/plots=surface(rotate=30 tilt=45);
run;
```

This gives Figure 6.11. There is no apparent advantage (and no apparent disadvantage) to the perspective plot over the contour plot in Figure 6.10 and which is used is essentially a matter of personal choice.

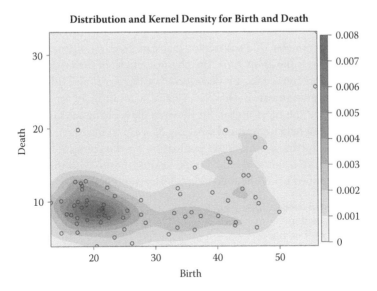

Figure 6.10 Scatterplot of birth and death rates for different countries enhanced by the contours of the estimated bivariate density of the two variables.

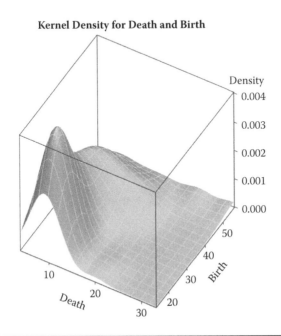

Figure 6.11 Perspective plot of birth and death rates for different countries.

6.7 US Birth Rates during and after World War II

An important parameter of a scatterplot that can greatly influence our ability to recognize patterns is the *aspect ratio*, the physical length of the vertical axis divided by that of the horizontal axis. By default SAS scales plots, and other graphics, to fill the available graphics area. This will, typically, result in an aspect ratio of 3:4, which may not be the most useful. To illustrate how changing this characteristic of a scatterplot can help understand what the data are trying to tell us, we shall use the monthly US births, per thousand population for the years 1940 to 1948 given in Table 6.5. A scatterplot of the birth rates against month with the default aspect ratio can be obtained using the following SAS instructions:

```
ods graphics/height=480 width=640;
proc sgplot data=usbirth;
  scatter y=rate x=obsdate;
  format obsdate year.;
run;
```

The units of height and width set on the **ods graphics** statement are in pixels, but other units could be used. The resulting plot is shown in Figure 6.12.

The plot shows that the US birth rate was increasing between 1940 and 1943, decreasing between 1943 and 1946, rapidly increasing during 1946, and then decreasing again during 1947 and 1948. As Cook and Weisberg (1982) comment:

> These trends seem to deliver an interesting history lesson since the U.S. involvement in World War II started in 1942 and troops began returning home during the part of 1945, about nine months before the rapid increase in the birth rate.

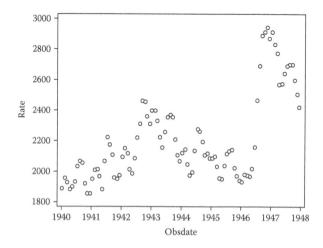

Figure 6.12   US birthrate against year with default aspect ratio.

Now let us see what happens when we alter the aspect ratio of the plot. In the following example we produce a plot with an aspect ratio of .3:

```
ods graphics/height=300 width=1000;
proc sgplot data=usbirth;
  scatter y=rate x=obsdate;
  format obsdate year.;
run;
```

The resulting graph appears in Figure 6.13. The new plot displays many peaks and troughs and suggests perhaps some minor within-year trends in addition to the global trends apparent in Figure 6.12. A clearer picture is obtained by plotting only a part of the data; here we will plot the observations for the years 1940 to 1943 using the SAS code:

```
proc sgplot data=usbirth;
  scatter y=rate x=obsdate;
  format obsdate monyy7.;
  where year<1943;
run;
```

The aspect ratio was set by the previous ODS graphics statement and will remain in force until reset or changed. This plot is shown in Figure 6.14.

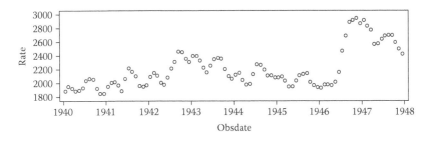

Figure 6.13   US birthrate against year with aspect ratio 0.3.

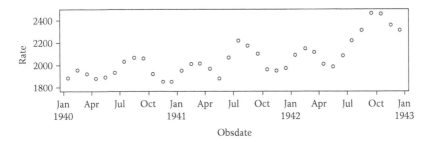

Figure 6.14   US birthrate against year (1940–1943) with aspect ratio 0.3.

Now, a within-year cycle is clearly apparent with the lowest within-year birth rate at the beginning of the summer and the highest occurring in the autumn. This pattern can be made clearer by connecting adjacent points in the plot with a line. The necessary SAS instructions are

```
proc sgplot data=usbirth;
  series y=rate x=obsdate;
  format obsdate monyy7.;
  where year<1943;
run;
```

The new plot appears in Figure 6.15. By reducing the aspect ratio to 0.2, replotting all 96 observations and again joining adjacent points with a line, both the within-year and global trends become clearly visible. The relevant SAS code is

```
ods graphics/height=200 width=1000;
proc sgplot data=usbirth;
  series y=rate x=obsdate;
  format obsdate year.;
run;
```

The plot appears in Figure 6.16.

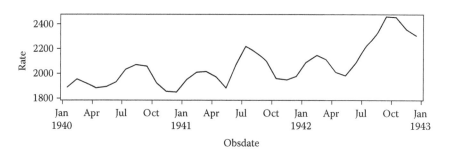

Figure 6.15 **US birthrate against year (1940–1943) with observations joined and aspect ratio 0.3.**

Figure 6.16 **US birthrate against year with observations joined and aspect ratio 0.2.**

6.8 Air Pollution in the United States

With the air pollution data it is likely that first thoughts for examining how the six ecological and climate variables determine sulphur dioxide (SO_2) concentration would be to fit a *multiple linear regression model*.

Multiple linear regression represents a generalization, to more than a single explanatory variable, of the simple linear regression procedure discussed previously in this chapter. It is now that the relationship between a response variable and several explanatory variables becomes of interest. Details of how to fit such models using SAS are given in Der and Everitt (2013). But as always in analysing a dataset some suitable graphic should precede any attempt to fit a model and an initial stage in coming to grips with this dataset would be to construct the scatterplots of each pair of variables in the data. For the air pollution data this would lead to twenty-one plots. Extracting the information in such a large number of separate plots would be difficult without turning them into what is known as a *scatterplot matrix* or sometimes a *draughtsman's plot*; this is nothing more than a square symmetric grid of all the bivariate scatterplots. When there are *p* variables in the dataset the grid has *p* rows and *p* columns, each one corresponding to a different variable. Each of the grid's cells shows a scatterplot of two variables. Variable *j* is plotted against variable *i* in the *ij*th cell, and the same two variables appear in cell *ji* with the *x*- and *y*-axes of the scatterplot interchanged. The reason for including both the upper and lower triangles of the grid, despite the seeming redundancy, is that it enables a row and column to be visually scanned to see one variable against all others, with the scales for the one variable lined up along the horizontal or vertical. As a result we can visually link features on one scatterplot with features on another and this ability greatly increases the power of the graphic.

So to begin we construct the scatterplot matrix for the seven variables in Table 6.6. This can be done by using the following code:

```
data usair;
    infile "c: \ hosgus \ data \ usair.dat";
    input city $16. so2 temperature factories population windspeed rain rainydays;
run;

proc sgscatter data=usair;
    matrix so2 -- rainydays;
run;
```

We have used **proc sgscatter**, which produces a range of plots that combine multiple variables. For a scatterplot matrix the **matrix** statement is used with a list of the variables to be included. There are only a few options for the **matrix** statement. The most useful is **diagonal=(***graph-list***)**, where *graph-list* is one or more of **histogram**, **kernel** and **normal** for histograms and kernel or normal density estimates, which are added to the diagonal cells instead of the variable names. A straightforward scatterplot matrix like the one here is also produced automatically by **proc corr**.

The result is shown in Figure 6.17. In this example where most interest centres on how air pollution as measured by sulphur dioxide concentration is related to the other six variables the scatterplots along the top row are perhaps of prime importance. Examining these six scatterplots in Figure 6.17 we see first that there are several outliers. For example in the plot of sulphur dioxide concentration against the 'factories' variable there is one city that has very high values on both these variables. Examination of the data in Figure 6.17 shows that this city is Chicago, which has a value of 110 micrograms per cubic metre for SO_2 concentration and 3344 manufacturing enterprises employing 20 or more workers. And because of its very high values of the indicator of air pollution, Chicago appears an outlier in the remaining five plots in the first row of Figure 6.17. The other clear outlier city in the six plots of sulphur dioxide concentration against the other variables is Providence with SO_2 concentration of 94 micrograms per cubic metre. The presence of these two very obvious

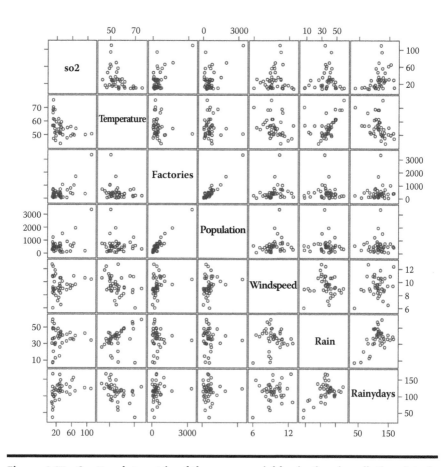

Figure 6.17   Scatterplot matrix of the seven variables in the air pollution data in Table 6.6.

outlier cities in the scatterplot matrix somewhat 'squashes' the points representing the remaining cities and makes the relationships between pairs of variables less clear than they might be. Consequently we shall replot the scatterplot matrix of the data now omitting Chicago and Providence, which can be achieved by adding the following where statement to the previous step:

```
where city not in('Chicago','Providence');
```

As usual with character variables, care is needed to match the case. A safer alternative would be

```
where lowcase(city) not in('chicago','providence');
```

The new scatterplot is shown in Figure 6.18.

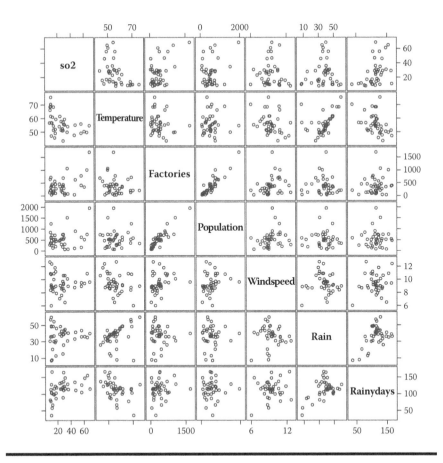

Figure 6.18   Scatterplot matrix of the air pollution data after removing the observations on Chicago and Providence.

Looking first at the first row of Figure 6.18 suggests that the relationships between the response variable, SO_2 concentration, and the six explanatory variables may not be linear suggesting something more complex than a straightforward multiple linear regression model may be needed to relate SO_2 concentration to the explanatory variables. We shall consider some possibilities later in the chapter.

Figure 6.18 also shows how the explanatory variables are related to each other. 'Strong' relationships between any pairs of explanatory variables raises questions about employing the usual multiple regression model on the data because highly correlated explanatory variables, that is *multicollinearity*, can cause several problems, for example:

1. It severely limits the size of the multiple correlation coefficient, R, because the explanatory variables are largely attempting to explain much of the same variability in the response variable (see Dizney and Gromen, 1967, for an example).
2. It makes determining the importance of a given explanatory variable difficult because the effects of explanatory variables are confounded due to their intercorrelations.
3. It increases the variances of the regression coefficients, making using the predicted model for prediction less stable. The parameter estimates become unreliable.

Spotting multicollinearity among a set of explanatory variables may not be easy, but examination of the portion of the scatterplot matrix involving the explanatory variables may often be helpful. If we do this for Figure 6.18 we see that using both number of factories and population size in a regression model for the data might lead to problems because they are very strongly related. Assessing the relationships between the explanatory variables becomes easier and the scatterplot matrix becomes more helpful if we show say the linear fits for each pair of variables on the appropriate panel of the plot along with the corresponding loess fit. We can do this using the following code (again we are leaving out Chicago and Providence):

```
proc sgscatter data=usair;
    compare x=(so2 -- rainydays) y=(so2 -- rainydays)/reg loess;
    where city not in('Chicago','Providence');
run;
```

Since the matrix statement in proc sgscatter does not include options for adding fitted lines to the data, we have used the compare statement. This produces scatterplots of pairs of variables. Variables listed (in parentheses) with x=(*variable List*) are paired with those listed with y=(*variable List*) and a range of fits can be specified as options. Here we use the same list of variables for x and y, which gives the result we require albeit with the somewhat redundant comparisons of variables with themselves on the diagonal.

The result is shown in Figure 6.19. Again examining the first row of Figure 6.19 shows that wind speed and amount of rainfall, in particular, appear to have distinctly non-linear effect on SO_2 concentration.

So the graphs we have looked at suggest, first, that fitting formal models to assess the effect of the six explanatory variables on SO_2 concentration should only be attempted after the removal of the observations for Chicago and Providence and additionally should not include both the population size and number of factories. And the graphs also suggest that a straightforward multiple regression model fitted to the data is unlikely to catch the subtleties of the relationships between SO_2 concentration and some of the explanatory variables. But to begin we shall ignore this last warning from the graphs and fit a multiple linear model to the air pollution after excluding the observations for Providence and Chicago and also

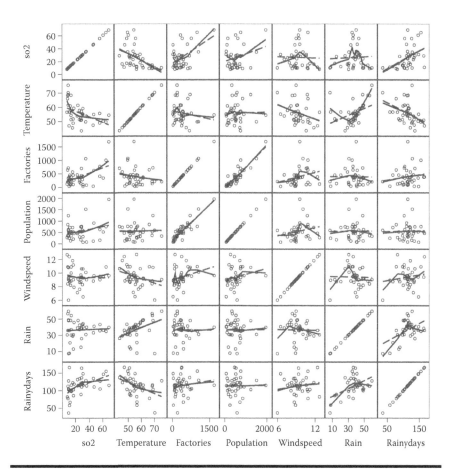

Figure 6.19   Scatterplot matrix of the air pollution data excluding Chicago and Providence, and showing linear and loess fits for each panel.

excluding population size from the list of explanatory variables. The SAS code to fit the model is

```
proc reg data=usair plots=residuals;
    model so2=temperature factories windspeed rain rainydays;
    where city not in('Chicago','Providence');
    output out=rout r=residual p=predicted;
run;
```

As well as producing a panel of residuals plotted against predictors via the plots=residuals option on the proc statement (result not shown), an output statement is included to save the residuals and predicted values to the dataset, rout, so that other plots can be produced as required.

An edited version of the numerical results is shown in Table 6.8.

It appears that the variables temperature, number of factories and wind speed are important for determining the level of air pollution in a city, with an increase in temperature conditional on the other variables leading to a decrease in pollution, an increase in number of factories conditional on the other variables leading to an increase in air pollution and higher wind speed conditional on the other variables leading to a decrease in air pollution. But we need to look at the residuals plotted against each of the five explanatory variables before we consider the results in Table 6.8 in any more detail. In general interpretation of these plots is helped by adding a locally weighted regression fit to each. To do this the saved residuals can be used as follows:

```
proc sgscatter data=rout;
  plot residual*(temperature factories windspeed rain rainydays)/loess;
run;
```

We have used proc sgscatter so that the plots are presented together in a panel, as given in Figure 6.20. For individual plots proc sgplot could have been used.

Table 6.8 Estimated Regression Coefficients, etc. for Multiple Linear Regression Model Fitted to Air Pollution Data

		Parameter Estimates			
Variable	DF	Parameter Estimate	Standard Error	t Value	Pr > |t|
Intercept	1	107.31942	39.52987	2.71	0.0105
temperature	1	−1.23964	0.50834	−2.44	0.0203
factories	1	0.02534	0.00632	4.01	0.0003
windspeed	1	−4.12074	1.53722	−2.68	0.0114
rain	1	0.31125	0.30995	1.00	0.3226
rainydays	1	0.05158	0.13862	0.37	0.7122

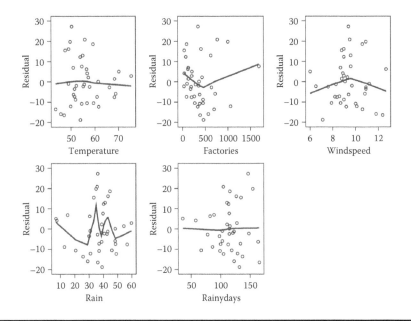

Figure 6.20   Residual plots with added locally weighted regression fits.

The residual plots for the variables, temperature, number of rainy days and number of factories (if we ignore the distortion caused by an outlier in this case) are essentially horizontal so we can, perhaps, assume that these variables are causing no problems in the model fitting process. But the loess fit on plot for wind speed shows some indication of 'smooth' curvature. (We shall ignore the rather less smooth loess fit for rainfall.) So now we will fit a second model to the data that allows for a quadratic effect of wind speed and construct the required residual plots corresponding to those in Figure 6.20. The necessary code is

```
data usair2;
  set usair;
  windspeed_q=windspeed*windspeed;
  if city not in('Chicago','Providence');
run;

proc reg data=usair2 plots=residuals;
  model so2=temperature factories windspeed windspeed_q rain rainydays;
  output out=rout2 r=residual p=predicted;
run;

proc sgscatter data=rout2;
  plot residual*(temperature factories windspeed windspeed_q rain rainydays)/
      loess;
run;
```

A short data step computes a new variable, windspeed_q, as the square of windspeed; this is needed as proc reg only allows existing variables to be used on the model statement. The data step also drops the observations for Chicago and Providence and stores the result in a new dataset, usair2. The proc reg step uses this dataset to refit the previous model with a quadratic function of windspeed, plots the residuals and saves them in the rout2 dataset.

The new residual plots are in Figure 6.21. The loess fit on the residual plot for wind speed now shows little departure from the horizontal suggesting that including the quadratic effect for wind speed may be sensible.

A panel plot showing the predicted values for the SO_2 values plotted against the observed values for the two models with added linear and loess fits can be constructed using the following SAS code:

```
data rout12;
  set rout rout2(in=in2);
  model=in2+1;
run;

proc sgpanel data=rout12;
  panelby model/columns=2 spacing=10;
  reg y=predicted x=so2;
  loess y=predicted x=so2/nomarkers;
run;
```

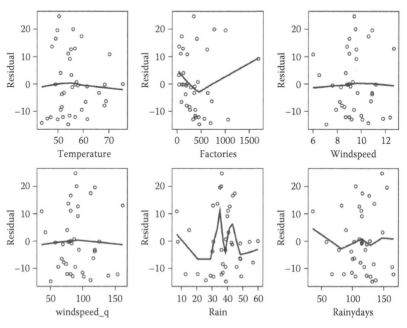

Figure 6.21 Residual plots for the second model fitted to the air pollution data.

First the two datasets containing the residuals from each model are combined into a new dataset with a variable, **model**, indicating which model the results come from, then they are plotted side by side using **proc sgpanel**. The **columns=2** option on the **panelby** statement ensures that they are displayed side by side rather than one above the other. Equivalent separate plots could have been produced using **proc sgplot** with each of the two datasets, but having them side by side with the axes uniformly scaled aids comparison.

The resulting graph is shown in Figure 6.22. Superficially the plots look very similar but closer examination of the one on the right-hand side of the diagram shows that two of the predicted SO_2 values are negative. Clearly the more complex model is not entirely sensible for the data despite our hopes for it from examining residual plots. One possibility would be to model $log(SO_2)$ values for constraining fitted SO_2 values to be greater than zero. This is left as an exercise for readers (see Exercise 6.6).

Once a suitable model has been decided on, it is often useful to illustrate the results graphically in the form of a plot of predicted values. Such a plot may be particularly helpful if the chosen model contains non-linear terms or interactions. Often the procedure used for the analysis will produce a suitable ODS plot. We saw in the previous chapter how **proc glm** can produce an interaction plot. In the case of the air pollution data we might want to show the relationship between SO_2 and wind speed

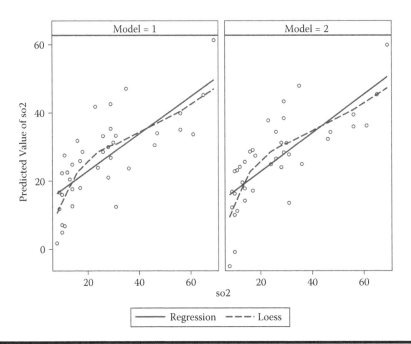

Figure 6.22 Plots of predicted SO_2 values against observed values for both models fitted to the air pollution data.

conditional on the other variables. A fit plot, like that shown in Figure 6.2, is not generally available as an ODS plot when there is more than one predictor variable in the model. To produce one a separate procedure, **proc plm**, is used as follows:

```
proc glm data=usair2;
  model so2=temperature factories windspeed I windspeed rain rainydays;
  store out=rmod;
run;
proc plm restore=rmod;
  effectplot fit(x=windspeed) / nocli;
run;
```

In order to use **proc plm** for a plot, the results of the model need to be stored in a special item store. Not all procedures have the facility to store model results in this way and in version 9.4 it is only partially implemented for **proc reg**, so we have used **proc glm** to re-run the model and store the results, via the **store** statement, in the rmod item store. Then **proc plm** accesses this item store and the **effectplot** statement is used to produce a fit plot with the *x*-axis variable, **windspeed**, specified in parentheses. By default the fitted values are calculated at the mean of the other predictors in the model, although the **at** option could be used to choose other specific values, if required. By default both confidence bands and prediction limits are plotted; the **nocli** option suppresses the latter. The result is shown in Figure 6.23.

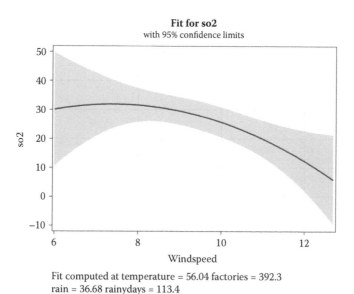

Fit computed at temperature = 56.04 factories = 392.3
rain = 36.68 rainydays = 113.4

Figure 6.23 Conditional fit plot for the model with quadratic effect of wind speed.

This suggests that wind speeds of up to 9 mph have little effect on SO_2 values but that they decline rapidly in stronger winds.

Given the difficulties of finding an entirely acceptable model for the air pollution data using a multiple regression approach, it might be productive to fit a *generalized additive model* as demonstrated in Der and Everitt (2013) as this enables the data themselves to suggest to the investigator in what way the response variable is related to each of the response variables without assuming linearity.

6.9 Summary

Simple and multiple linear regression are two of the most widely used (and misused) statistical techniques. Fitting such models should *always* be preceded by constructing suitable graphics for the data, which will avoid fitting such models to data for which they are clearly unsuitable. And when such models *are* fitted, graphical examination of residuals and other diagnostic quantities is essential to test assumptions and so on.

Exercises

6.1. Repeat the analysis of the galaxy velocity data after removing galaxies 3, 15 and 20 for the reasons given in the text. What is the estimated age of the universe from the new analysis?

6.2. The data in Table 6.9 give the chest, hips and waist measurements (in inches) of each of 20 individuals. Construct a bubble plot of the data. Does the plot reveal any 'pattern' in the data and if so can you give an explanation?

Table 6.9 Chest, Waist and Hips Measurement

Individual	Chest	Waist	Hips
1	34	30	32
2	37	32	37
3	38	30	36
4	36	33	39
5	38	29	33
6	43	32	38
7	40	33	42
8	38	30	40
9	40	30	37

Continued

Table 6.9 (*Continued*) Chest, Waist and Hips Measurement

Individual	Chest	Waist	Hips
10	41	32	39
11	36	24	35
12	36	25	37
13	34	24	37
14	33	22	34
15	36	26	38
16	37	26	37
17	34	25	38
18	36	26	37
19	38	28	40
20	35	23	35

6.3. The iron content of crushed blast furnace slag can be determined by a chemical test at a laboratory or estimated by a cheaper, quicker magnetic test. The data in Table 6.10 were collected to investigate the effect to which the results of the chemical test of iron content can be predicted from a magnetic test of iron content, and the nature of the relationship between these quantities. Construct a scatterplot of the data. What does this tell you about the relationship between the chemical and magnetic tests? The observations in Table 6.10 are given in the time order in which they were made. Does this have any effect?

Table 6.10 Iron in Slag

Chemical	Magnetic	Chemical	Magnetic
24	25	20	21
16	22	20	21
24	17	25	21
18	21	27	25
18	20	22	22
10	13	20	18

Table 6.10 (*Continued*)   Iron in Slag

Chemical	Magnetic	Chemical	Magnetic
14	16	24	21
16	14	24	18
18	19	23	20
20	10	29	25
21	23	27	20
20	20	23	18
21	19	19	19
15	15	25	16
16	16	15	16
15	16	16	16
17	12	27	26
19	15	27	28
16	15	30	28
15	15	29	30
15	15	26	32
13	17	25	28
24	18	25	36
22	16	32	40
21	18	28	33
24	22	25	33
15	20		

6.4. The Hertzsprung–Russell (H–R) diagram forms the basis of the theory of stellar evolution. The diagram is essentially a plot of the energy output of stars plotted against their surface temperature. Data from the star cluster CYG OB1, calibrated according to Vanisma and De Greve (1972), are shown in Table 6.11. Construct the scatterplot of the data and show on the plot the estimated bivariate probability density function of the two variables.

Table 6.11   Star Cluster CYG OB1

Index	Log Surface Temperature	Log Light Intensity	Index	Log Surface Temperature	Log Light Intensity
1	4.37	5.23	25	4.38	5.02
2	4.56	5.74	26	4.42	4.66
3	4.26	4.93	27	4.29	4.66
4	4.56	5.74	28	4.38	4.90
5	4.30	5.19	29	4.22	4.39
6	4.46	5.46	30	3.48	6.05
7	3.84	4.65	31	4.38	4.42
8	4.57	5.27	32	4.56	5.10
9	4.26	5.57	33	4.45	5.22
10	4.37	5.12	34	3.49	6.29
11	3.49	5.73	35	4.23	4.34
12	4.43	5.45	36	4.62	5.62
13	4.48	5.42	37	4.53	5.10
14	4.01	4.05	38	4.45	5.22
15	4.29	4.26	39	4.53	5.18
16	4.42	4.58	40	4.43	5.57
17	4.23	3.94	41	4.38	4.62
18	4.42	4.18	42	4.45	5.06
19	4.23	4.18	43	4.50	5.34
20	3.49	5.89	44	4.45	5.34
21	4.29	4.38	45	4.55	5.54
22	4.29	4.22	46	4.45	4.98
23	4.42	4.42	47	4.42	4.50
24	4.49	4.85			

6.5. Weather modification, or cloud seeding, is the treatment of individual clouds or storm systems with various inorganic or organic materials in the hope of achieving an increase in rainfall. Introduction of such material into a cloud that contains supercooled water, that is liquid water colder than zero degrees Celsius, has the aim of inducing freezing, with the consequent ice particles growing at the expense of liquid droplets and becoming heavy enough to fall as rain from clouds that otherwise would produce none.

The data shown in Table 6.12 were collected in the summer of 1975 from an experiment to investigate the use of massive amounts of silver iodide (100 to 1000 grams per cloud) in cloud seeding to increase rainfall (Woodley et al., 1977). In the experiment, which was conducted in an area of Florida, 24 days were judged suitable for seeding on the basis that a measured suitability criterion, denoted *S-Ne,* was not less than 1.5. Here S is the 'seedability', the difference between the maximum height of a cloud if seeded and the same cloud if not seeded predicted by a suitable cloud model, and Ne is the number of hours between 1300 and 1600 GMT with 10 centimetre echoes in the target; this quantity biases the decision for experimentation against naturally rainy days. Consequently, optimal days for seeding are those on which seedability is large and the natural rainfall early in the day is small.

Table 6.12 Cloud Seeding Data

Seeding	Time	S-Ne	Cloud Cover	Prewetness	Echomotion	Rainfall
0	0	1.75	13.40	0.274	2	12.85
1	1	2.70	37.90	1.267	1	5.52
1	3	4.10	3.90	0.198	2	6.29
0	4	2.35	5.30	0.526	1	6.11
1	6	4.25	7.10	0.250	1	2.45
0	9	1.60	6.90	0.018	2	3.61
0	18	1.30	4.60	0.307	1	0.47
0	25	3.35	4.90	0.194	1	4.56
0	27	2.85	12.10	0.751	1	6.35
1	28	2.20	5.20	0.084	1	5.06
1	29	4.40	4.10	0.236	1	2.76

Continued

Table 6.12 (*Continued*)   Cloud Seeding Data

Seeding	Time	S-Ne	Cloud Cover	Prewetness	Echomotion	Rainfall
1	32	3.10	2.80	0.214	1	4.05
0	33	3.95	6.80	0.796	1	5.74
1	35	2.90	3.00	0.124	1	4.84
1	38	2.05	7.00	0.144	1	11.86
0	39	4.00	11.30	0.398	1	4.45
0	53	3.35	4.20	0.237	2	3.66
1	55	3.70	3.30	0.960	1	4.22
0	56	3.80	2.20	0.230	1	1.16
1	59	3.40	6.50	0.142	2	5.45
1	65	3.15	3.10	0.073	1	2.02
0	68	3.15	2.60	0.136	1	0.82
1	82	4.01	8.30	0.123	1	1.09
0	83	4.65	7.40	0.168	1	0.28

On suitable days, a decision was taken at random as to whether to seed or not. For each day the following variables were measured:

Seeding – A factor indicating whether seeding action occurred (1 = yes or 0 = no).

Time – Number of days after the first day of the experiment.

S-Ne – Suitability criterion (see earlier).

Cloud cover – Percentage of cloud cover in the experimental area, measured using radar.

Prewetness – Total rainfall in the target area one hour before seeding (in cubic metres $\times$ 10[7]).

Echomotion – A factor showing whether the radar echo was moving or stationary.

Rainfall – Amount of rain in cubic metres $\times 10^7$.

The objective in analysing these data is to see how rainfall is related to the other variables and, in particular, to determine the effectiveness of seeding. Produce some suitable graphics for the data and use them to suggest a possible model for the data and then fit this model. What are your conclusions?

6.6. Reanalyse the air pollution data using $\log(SO_2)$ as the response variable.

Chapter 7

Graphs for Logistic Regression: Blood Screening, Women's Role in Society and Feeding Alligators

7.1 Introduction

In this chapter we will be concerned with graphics which can be helpful when applying logistic regression, a form of analysis used when a response variable is categorical, most commonly with two categories, but at times more than two. The following three sets of data will be used in the chapter:

Blood screening—The erythrocyte sedimentation rate (ESR) is the rate at which red blood cells (erythrocytes) settle out of suspension in blood plasma when measured under standard conditions. If the ESR increases when the level of certain proteins in the blood plasma rise in association with conditions such as rheumatic disease, chronic infections and malignant diseases, its determination might be useful in screening blood samples taken from people suspected of suffering from one of the conditions mentioned. The absolute value of the ESR is not of great importance;

Table 7.1 Blood Plasma Data

Fibrinogen	Gamma Globulin	ESR = >20	Fibrinogen	Gamma Globulin	ESR = >20
2.52	38	0	2.88	30	0
2.56	31	0	2.65	46	0
2.19	33	0	2.28	36	0
2.18	31	0	2.67	39	0
3.41	37	0	2.29	31	0
2.46	36	0	2.15	31	0
3.22	38	0	2.54	28	0
2.21	37	0	3.34	30	0
3.15	39	0	2.99	36	0
2.60	41	0	3.32	35	0
2.29	36	0	5.06	37	1
2.35	29	0	3.34	32	1
3.15	36	0	2.38	37	1
2.68	34	0	3.53	46	1
2.60	38	0	2.09	44	1
2.23	37	0	3.93	32	1

rather, less than 20 mm/hr indicates a 'healthy' individual. To assess whether or not the ESR is a useful diagnostic tool, Collett and Jemain (1985) collected the data shown in Table 7.1. The question of interest is whether there is any association between the probability of an ESR reading greater than 20 mm/hr and the level of the two plasma proteins. If there is not, then the determination of ESR would not be useful for diagnostic purposes.

Women's role in society—In a survey carried out in the 1970s each respondent was asked if he or she agreed or disagreed with the statement, 'Women should take care of running their home and leave running the country up to men'. The responses are summarized in Table 7.2 (the data are given in Haberman, 1972, and Collett, 2003). The questions of interest here are whether the response of men and women differ and how years of education affect the response.

Table 7.2 Women's Role in Society Data

Education	Gender	Agree	Disagree	Education	Gender	Agree	Disagree
0	M	4	2	0	F	4	2
1	M	2	0	1	F	1	0
2	M	4	0	2	F	0	0
3	M	6	3	3	F	6	1
4	M	5	5	4	F	10	0
5	M	13	7	5	F	14	7
6	M	25	9	6	F	17	5
7	M	27	15	7	F	26	16
8	M	75	49	8	F	91	36
9	M	29	29	9	F	30	35
10	M	32	45	10	F	55	67
11	M	36	59	11	F	50	62
12	M	115	245	12	F	190	403
13	M	31	70	13	F	17	92
14	M	28	79	14	F	18	81
15	M	9	23	15	F	7	34
16	M	15	110	16	F	13	115
17	M	3	29	17	F	3	28
18	M	1	28	18	F	0	21
19	M	2	13	19	F	1	2
20	M	3	20	20	F	2	4

Feeding alligators—Agresti (1996) reports a study undertaken by the Florida Game and Fresh Water Fish Commission of factors influencing the primary food choice of alligators. For 59 alligators sampled in Lake George, Florida, Table 7.3 shows the alligator's length (in metres) and the primary food type (by volume) found in the alligator's stomach. Primary food type has three categories: Fish (F), Invertebrate (I) and Other (O). The invertebrates were primarily apple snails, aquatic insects and crayfish. The 'other' category included reptiles (primarily turtles, though one stomach contained tags of 23 baby alligators that had been released in the lake during the previous year!), amphibians,

Table 7.3 Alligator Length and Primary Food Choice

Length	Choice	Length	Choice	Length	Choice	Length	Choice
1.30	I	1.80	F	1.24	I	2.56	O
1.32	F	1.85	F	1.30	I	2.67	F
1.32	F	1.93	I	1.45	I	2.72	I
1.40	F	1.93	F	1.45	O	2.79	F
1.42	I	1.98	I	1.55	I	2.84	F
1.42	F	2.03	F	1.60	I		
1.47	I	2.03	F	1.60	I		
1.47	F	2.31	F	1.65	F		
1.50	I	2.36	F	1.78	I		
1.52	I	2.46	F	1.78	O		
1.63	I	3.25	O	1.80	I		
1.65	O	3.28	O	1.88	I		
1.65	O	3.33	F	2.16	F		
1.65	I	3.56	F	2.26	F		
1.65	F	3.58	F	2.31	F		
1.68	F	3.66	F	2.36	F		
1.70	I	3.68	O	2.39	F		
1.73	O	3.71	F	2.41	F		
1.78	F	3.89	F	2.44	F		
1.78	O						

Note: F, fish; I, invertebrate; O, other.

mammals, plant material, and stones or other debris. The question of interest is how/if length of an alligator predicts food choice.

7.2 Logistic Regression

Multiple linear regression was discussed in the previous chapter. One way to write the model underlying this method is as $y \sim N(\mu, \sigma^2)$ where $\mu = \beta_0 + \beta_1 x_1 + \ldots + \beta_q x_q$. This makes it clear that this model is suitable for continuous response variables with, conditional on the values of the explanatory variables, a normal distribution with constant variance. So clearly the model would not be suitable for the three datasets described in the introduction where the response variable is categorical, with for the first two datasets, two categories which would be labelled 0 and 1, and for the third dataset, three categories which would be labelled 1, 2 and 3. If we were to model the expected value of, for example, the binary response, i.e. the probability that it takes the value one, say π, directly as a linear function of explanatory variables, it could lead to fitted values of the response probability outside the range [0,1], which clearly would not be sensible. So instead of modelling the expected value of the response directly as a linear function of explanatory variables, a suitable transformation is so modelled. For a binary response the *logit* function is used to give the model

$$\text{logit}(\pi) = \log\left(\frac{\pi}{1-\pi}\right) = \beta_0 + \beta_1 x_1 + \ldots \beta_q x_q \tag{7.1}$$

(We have used the same terms for the regression coefficients in Equation 7.1 as earlier for the multiple linear regression model for convenience. They are not, of course, the same values.)

The logit of a probability is simply the log of the odds of the response taking the value one. Equation 7.1 can be rewritten as

$$\pi = \frac{\exp(\beta_0 + \beta_1 x_1 + \ldots \beta_q x_q)}{1 + \exp(\beta_0 + \beta_1 x_1 + \ldots \beta_q x_q)} \tag{7.2}$$

Parameters in the model are estimated by maximum likelihood; details are given in Collett (2003).

Logit models for categorical response variables with more than two categories pair each response category with a baseline category, the choice of which is arbitrary. So if we have a response variable with, say, K categories, we might use the last category as the baseline, in which case our model becomes (using an obvious nomenclature and restricting ourselves to the case when there is a single explanatory variable)

$$\log\left(\frac{\pi_k}{\pi_K}\right) = \alpha_k + \beta_k x, \quad k = 1, 2 \ldots K-1 \tag{7.3}$$

The model consists of $K-1$ logit equations, with separate parameters for each. When $K = 2$ the model becomes that described earlier for a binary response. Parameters are estimated by fitting the $K-1$ logit equations simultaneously. The model in Equation 7.3 can be alternatively written in terms of response probabilities as follows:

$$\pi_k = \frac{\exp(\alpha_j + \beta_j x)}{\sum_h \exp(\alpha_h + \beta_h x)}, \quad k = 1, \ldots, K-1 \tag{7.4}$$

For more details of logistic regression models, see Hosmer, Lemeshow and Sturdivant (2013).

7.3 Blood Screening

For the blood screening data the response variable, ESR, takes just two values, which are determined by whether it is above 20 mm/hr. We will fit a logistic regression model with the two explanatory variables: fibrinogen and gamma globulin. How this is done in SAS is given in Der and Everitt (2013). The edited results are shown in Table 7.4.

The results strongly suggest that only the fibrinogen value is really useful for predicting whether or not the ESR will be greater than 20 mm/hr. But it is useful to examine some graphics and here we will first plot the predicted probabilities of an ESR values greater than 20 mm/hr against the values of each explanatory variable. To do this we use the **effectplot** statement within **proc logistic**.

```
proc format;
    value esr 0='<20' 1='>20';
run;
proc logistic data=plasma desc;
    model esr=fibrinogen gamma;
    effectplot fit (x=gamma);
    effectplot fit (x=fibrinogen);
    output out=esrout predicted=pred;
    format esr esr.;
run;
```

The **effectplot** statement is useful for plotting the fitted values from models which are too complex for the **plots=** option on the proc statement. In this case, **plots=effect** would only have plotted the first predictor, **fibrinogen**, against the predicted probabilities. On the **effectplot** statement, the variable against which the predicted probabilities are plotted is specified in the parentheses with the **x=** option. By default, the fitted values for display in the plot are calculated at the mean of any other continuous predictors in the model and/or the reference level of any categorical predictors. Here two **effectplot** statements are used to plot the fitted values against each predictor at the mean of the other.

The two plots are shown in Figures 7.1 and 7.2.

Figure 7.2 clearly shows the greater predictive ability of fibrinogen over gamma globulin but the confidence interval for prediction shown by the shaded area in the figure shows that the prediction is far from accurate particularly for

Table 7.4   Results from Logistic Regression of Blood Plasma Data

Parameter	DF	Estimate	Standard Error	Wald Chi-Square	Pr > ChiSq
Intercept	1	−12.7920	5.7964	4.8704	0.0273
Fibrinogen	1	1.9104	0.9710	3.8708	0.0491
Gamma	1	0.1558	0.1195	1.6982	0.1925

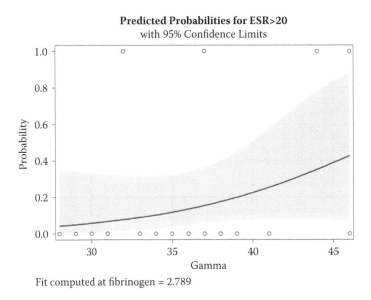

Figure 7.1 **Predicted probability of an ESR value equal to or greater than 20 against the values of the gamma globulin explanatory variable with fibrinogen fixed at 2.789.**

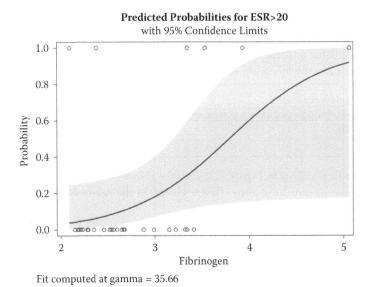

Figure 7.2 **Predicted probability of an ESR value equal to or greater than 20 against values of fibrinogen with the value of gamma globulin fixed at 35.66.**

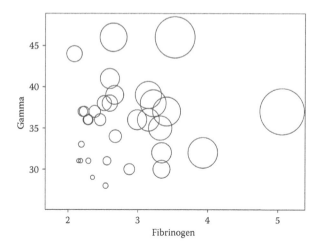

Figure 7.3 **Bubble plot of fitted probabilities of an ESR value equal to or greater than 20 from a fitted logistic regression model against the values of the two explanatory variables in the model.**

higher probabilities due to the relatively small number of values of ESR greater than 20 mm/hr.

Although it seems in this example one of the explanatory variables, gamma globulin, is not of great use for predicting the ESR value, it is of interest to plot the predicted values from the logistic model containing both explanatory variables using a bubble plot (see Chapter 6); in this plot the radius of the circles will represent the predicted probabilities of an ESR value equal to or greater than 20 mm/hr. For this we need the following code:

```
proc sgplot data=esrout;
  bubble x=fibrinogen y=gamma size=pred/bradiusmin=1pct bradiusmax=
  10pct nofill;
run;
```

The result is shown in Figure 7.3. The plot shows the increasing probability of an ESR value 20 mm/hr or above (larger circles) as the values of fibrinogen and, to a lesser extent, gamma globulin increase.

7.4 Women's Role in Society

To begin with the data from women's role in society we will fit a logistic regression model that includes both explanatory variables, gender and years of education, to the data. Again, for details of how to do this in SAS, see Der and Everitt (2013). The results of the model fitting are shown in Table 7.5.

Table 7.5   Results from Logistic Regression for the Women's Role in Society Data

Analysis of Maximum Likelihood Estimates					
Parameter	DF	Estimate	Standard Error	Wald Chi-Square	Pr > ChiSq
Intercept	1	2.5093	0.1839	186.2102	<0.0001
Education	1	−0.2706	0.0154	308.3258	<0.0001
Gender F	1	−0.0115	0.0841	0.0185	0.8917

The results shown in Table 7.5 appear to show that years of education has a highly significant part to play in determining how a person will respond to the statement read to them, with the person's gender apparently unimportant. The relationship between the probability of agreeing with the statement and years of education can be illustrated graphically by a diagram showing the observed proportions of agreeing for men and women with those predicted by the model plotted against years of education. The required plot can be constructed using the following SAS code:

```
proc logistic data=roles;
   class gender/param=ref;
   model agree/total=education gender;
   effectplot slicefit (x=education sliceby=gender);
run;
```

The result is shown in Figure 7.4. The two curves for males and females are almost identical reflecting the non-significant value of the regression coefficient for gender in the fitted logistic regression model. But the observed values plotted on Figure 7.4 suggest that there could be an interaction between years of education and gender, a possibility that can be investigated by applying a further logistic regression model, the results of which are shown in Table 7.6.

The interaction between gender and years of education is seen to be highly significant. Interpreting the regression coefficient for the interaction is made simpler if again we plot the observed proportions and those predicted by the new model using the following code:

```
proc logistic data=roles;
   class gender/param=ref;
   model agree/total=education|gender;
   effectplot slicefit (x=education sliceby=gender)/clm;
run;
```

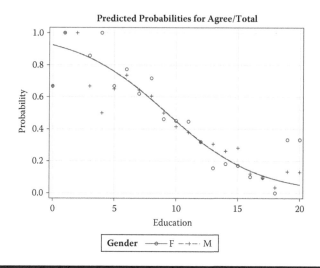

Figure 7.4 Fitted and observed values for probability of agreeing from a logistic model containing years of education and gender for the women's role in society data.

Table 7.6 Results from Fitting a Logistic Regression Model Including Years of Education, Gender and the Years × Gender Interaction to the Women's Role in Society Data

Analysis of Maximum Likelihood Estimates					
Parameter		DF	Estimate	Standard Error	Wald Chi-Square	Pr > ChiSq
Intercept		1	2.0982	0.2355	79.3787	<.0001
Education		1	−0.2340	0.0202	134.3777	<.0001
Gender	F	1	0.9046	0.3601	6.3117	0.0120
Education*Gender	F	1	−0.0814	0.0311	6.8490	0.0089

The result appears in Figure 7.5. We see that for fewer years of education women have, a higher probability of agreeing with the statement than men, but when the years of education exceed about 10 then this situation reverses.

It is also helpful to graphically display the numerical results from fitting the interaction model by plotting the various odds ratios and their confidence intervals. Such a plot can be produced by including an appropriate oddsratio statement in the proc logistic step. For example, including

```
oddsratio gender/at(education=5 10 15);
```

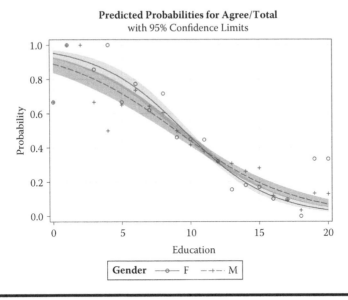

Figure 7.5 **Observed and fitted values for probability of agreeing when a logistic regression model with years of education, gender and the years × gender interaction as explanatory variables is fitted to the women's role in society data.**

in the previous step results in the plot shown in Figure 7.6. This plot shows that the odds in favour of agreeing with the statement are greater for women than for men at the 5-year education level, with no difference for 10 years of education and odds greater for men than for women for 15 years of education.

A range of residuals and other diagnostics is available for use in association with logistic regression to check whether particular components of the model are adequate. A comprehensive account of these is given in Collett (2003). Here we shall look at only one of the possible residuals, namely, the one known as the *deviance residual*. The residual is defined explicitly in Der and Everitt (2013) and it provides information about how well a model fits each particular observation when it is plotted against fitted value. For a 'good' model the residuals should all (or at least the great majority) lie within a horizontal band between -2 and 2. We can construct the required plot for the gender × years of education model as follows:

```
proc sgplot data=rolesout;
   scatter y=dres x=pred;
   refline 2 -2/axis=y lineattrs=(pattern=dash);
run;
```

The result is shown in Figure 7.7. The residuals largely fall into the $[-2,2]$ range and the plot does not suggest a poor fit for any particular observation or subset of observations.

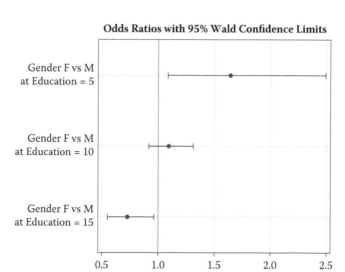

Figure 7.6   Odds ratios and 95% CIs for a gender difference at three different levels of education.

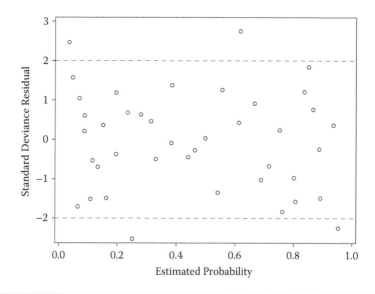

Figure 7.7   Plot of deviance residuals from logistic regression models containing years of education, gender and years × gender interaction fitted to the women's role in society data.

7.5 Alligator Food

We can apply the logit model described in Section 7.2 for a categorical response (food type) with K = 3 categories to the alligator data using the following code:

```
proc logistic data=alligators;
  model food (ref='O')=length/link=glogit;
  effectplot fit (x=length);
  oddsratio length;
run;
```

(Note that food type 'Other' is used as the baseline category.)

The numerical results are shown in Table 7.7.

The parameter estimates in this model contrast 'Fish' and 'Invertebrate' to 'Other' as the baseline category. Remembering that the intercept and slope parameter for 'Other' are both set at zero we can, using Equation 7.4, estimate the probabilities of the outcomes, 'Fish', 'Invertebrate', 'Other' as follows:

$$\pi_1 = \frac{\exp(0.99 + 0.08\ \text{length})}{1 + \exp(0.99 + 0.08\ \text{length}) + \exp(5.18 - 2.39\ \text{length})}$$

$$\pi_2 = \frac{\exp(5.18 - 2.39\ \text{length})}{1 + \exp(0.99 + 0.08\ \text{length}) + \exp(5.18 - 2.39\ \text{length})}$$

$$\pi_3 = \frac{1}{1 + \exp(0.99 + 0.08\ \text{length}) + \exp(5.18 - 2.39\ \text{length})}$$

Table 7.7   Parameter Estimates for Logit Model Fitted to Alligator Feeding Data

Analysis of Maximum Likelihood Estimates						
Parameter	Food	DF	Estimate	Standard Error	Wald Chi-Square	Pr > ChiSq
Intercept	F	1	0.9985	1.1763	0.7206	0.3960
Intercept	I	1	5.1809	1.7458	8.8069	0.0030
Length	F	1	0.0848	0.4885	0.0301	0.8622
Length	I	1	−2.3884	0.9215	6.7182	0.0095

Odds Ratio Estimates and Wald Confidence Intervals		
Label	Estimate	95% Confidence Limits
Food F: length	1.089	0.418	2.836
Food I: length	0.092	0.015	0.559

The **effectplot** statement produces the plot shown in Figure 7.8, which graphs these three response probabilities as a function of length. Clearly as the length of the alligator increases there is an increased probability of fish being on the alligator's menu. (There remains the possibility that the sex of the alligator may affect diet and Exercise 7.3 will allow you to investigate this possibility.)

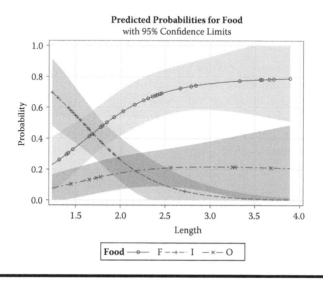

Figure 7.8 Predicted probabilities of alligator's food choice plotted against length of alligator.

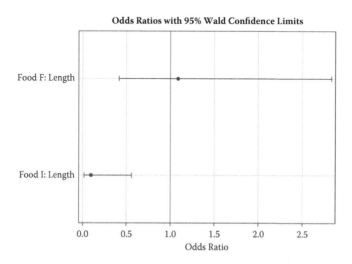

Figure 7.9 Odds ratio for food type per 1 metre increase in length of alligator.

The **oddsratio** statement included in the **proc logistic** step gives the results both in terms of the odds ratios as shown in Table 7.7 and as a graph shown in Figure 7.9. In both cases the odds ratios are per unit increase in the length of the alligator, that is, for alligators a metre different in length.

7.6 Summary

Logistic regression is a widely used piece of statistical methodology. Interpretation of the results from fitting such models is almost always made clearer by using some of the graphics described in this chapter.

Exercises

7.1. Agresti (1996) gives the data shown in Table 7.8 relating to the tragic flight of the space shuttle *Challenger* (see Chapter 1). (The data were given originally in Dalal et al., 1989.) Fit a logistic regression model to the data and use the results in association with some informative graphic to investigate the effect of temperature on the probability of thermal distress.

Table 7.8 Temperature (F) at the Time of Flight and Whether at Least One Primary O-Ring Suffered Thermal Distress

Flight	Temp	TD	Flight	Temp	TD
1	66	0	13	67	0
2	70	1	14	53	1
3	69	0	15	67	0
4	68	0	16	75	0
5	67	0	17	70	0
6	72	0	18	81	0
7	73	0	19	76	0
8	70	0	20	79	0
9	57	1	21	75	1
10	63	1	22	76	0
11	70	1	23	58	1
12	78	0			

Note: TD, thermal distress (1 = yes, 0 = no).

7.2. The data in Table 7.9 arise from a study of a psychiatric screening questionnaire called the General Health Questionnaire (GHQ) described in Goldberg (1972). Fit logistic regression models and use suitable graphics to investigate how 'caseness' is related to gender and GHQ score.

Table 7.9 Psychiatric Screening Data

GHQ Score	Sex	Number of Cases	Number of Non-Cases
0	F	4	80
1	F	4	29
2	F	8	15
3	F	6	3
4	F	4	2
5	F	6	1
6	F	3	1
7	F	2	0
8	F	3	0
9	F	2	0
10	F	1	0
0	M	1	36
1	M	2	25
2	M	2	8
3	M	1	4
4	M	3	1
5	M	3	1
6	M	2	1
7	M	4	2
8	M	3	1
9	M	2	0
10	M	2	0

Note: F, female; M, male.

7.3. The data in Table 7.10 are the same as in Table 7.3 but with the addition of the gender of each alligator. Reanalyse the data using no both length and gender as explanatory variables and illustrate the results with suitable graphics. What conclusions can you draw about the data?

Table 7.10   Alligator Feeding Data with the Gender of Each Alligator Identified

Males				Females			
Length	*Choice*	*Length*	*Choice*	*Length*	*Choice*	*Length*	*Choice*
1.30	I	1.80	F	1.24	I	2.56	O
1.32	F	1.85	F	1.30	I	2.67	F
1.32	F	1.93	I	1.45	I	2.72	I
1.40	F	1.93	F	1.45	O	2.79	F
1.42	I	1.98	I	1.55	I	2.84	F
1.42	F	2.03	F	1.60	I		
1.47	I	2.03	F	1.60	I		
1.47	F	2.31	F	1.65	F		
1.50	I	2.36	F	1.78	I		
1.52	I	2.46	F	1.78	O		
1.63	I	3.25	O	1.80	I		
1.65	O	3.28	O	1.88	I		
1.65	O	3.33	F	2.16	F		
1.65	I	3.56	F	2.26	F		
1.65	F	3.58	F	2.31	F		
1.68	F	3.66	F	2.36	F		
1.70	I	3.68	O	2.39	F		
1.73	O	3.71	F	2.41	F		
1.78	F	3.89	F	2.44	F		
1.78	O						

Note: F, fish; I, invertebrate; O, other.

Chapter 8

Graphing Longitudinal Data: Glucose Tolerance Tests and Cognitive Behavioural Therapy (CBT) for Depression

8.1 Introduction

Two sets of data will be of interest in this chapter. Both of them involve *longitudinal data*, which arise when the participants in a study have the value of some variable of interest measured or observed on several different occasions. Such data are often said to contain *repeated measurements* of the response variable of interest.

> *Glucose tolerance tests*—In a study of the association between hyperglycemia and relative hyperinsulinemia, standard glucose tolerance tests were administered to three groups of subjects: 13 controls, 12 non-hyperinsulinemic obese patients and 8 hyperinsulinemic patients (see Zerbe, 1979, and Zerbe and Murphy, 1986). Plasma inorganic phosphate (PIP) measurements were obtained from blood samples taken 0, 0.5, 1, 1.5, 2, 3, 4 and 5 hours after a standard-dose oral glucose challenge. Part of the data is given in Table 8.1. The question of interest: Do the three groups differ in their responses to the glucose challenge test?

Table 8.1 Sub-Set of the Results from Oral Glucose Challenge for Three Groups of Subjects

		Hours after Glucose Challenge							
Group	Id No	0	0.5	1	1.5	2	3	4	5
Control	1	4.3	3.3	3.0	2.6	2.2	2.5	3.4	4.4
	2	3.7	2.6	2.6	1.9	2.9	3.2	3.1	3.9
	3	4.0	4.1	3.1	2.3	2.9	3.1	3.9	4.0
...									
	13	4.7	3.1	3.2	3.3	3.2	4.2	3.7	4.3
Non-hyperinsulinemic	14	4.3	3.3	3.0	2.6	2.2	2.5	2.4	3.4
	15	5.0	4.9	4.1	3.7	3.7	4.1	4.7	4.9
	16	4.6	4.4	3.9	3.9	3.7	4.2	4.8	5.0
...									
Hyperinsulinemic	26	4.9	4.3	4.0	4.0	3.3	4.1	4.2	4.3
	27	5.1	4.1	4.6	4.1	3.4	4.2	4.4	4.9
	28	4.8	4.6	4.6	4.4	4.1	4.0	3.8	3.8
...									

Cognitive behavioural therapy (CBT) for depression—The design of most longitudinal studies specifies that all participants in the study are to have their measurements of the response variable made at a common set of time points. But often intended measurements are not made when, for example, staff fail to make a scheduled observation or when the participant simply fails to turn up for an appointment. In some cases a patient may drop out of the study altogether at some point. An example of a longitudinal study (a clinical trial in this case) in which some patients drop out is one described in Proudfoot et al. (2003) in which a computerized method of delivering cognitive behavioural therapy for depression known as 'Beating the Blues (BtB)' was assessed against 'Treatment as Usual (TAU)' the latter whatever the general practitioner offered with the exception of any face-to-face counselling of psychological intervention. The response variable here was the Beck Depression Inventory II (BDI; Beck et al., 1996), which was supposed to be recorded on five occasions, prior to treatments and at 2, 4, 6 and 8 months after treatment began. Data for the first 10 patients are shown in Table 8.2.

Table 8.2   Data for Ten Patients in the BtB Trial

Sub	Treatment	BDIpre	BDI2m	BDI4m	BDI6m	BDI8m
1	TAU	29	2	2	.	.
2	BtB	32	16	24	17	20
3	TAU	25	20	.	.	.
4	BtB	21	17	16	10	9
5	BtB	26	23	.	.	.
6	BtB	7	0	0	0	0
7	TAU	17	7	7	3	7
8	TAU	20	20	21	19	13
9	BtB	18	13	14	20	11
10	BtB	20	5	5	8	12

The periods in Table 8.2 indicate whether a protocol-specified measurement of the BDI was not made. Here all the missing values are due to patients dropping out of the study. How dropouts might affect the results of formal analyses of the data will be discussed later in the chapter.

8.2 Graphical Displays for Longitudinal Data

Graphical displays of longitudinal data can be useful for getting an overall picture of the data before any formal analyses are carried out and also for informing such formal analyses in the sense of pointing to models that are likely to be the most sensible for the data. According to Diggle et al. (2002), there is no single prescription for making effective graphical displays of longitudinal data, although they do offer the following simple guidelines:

- Show as much of the relevant raw data as possible rather than only data summaries.
- Highlight aggregate patterns of potential scientific interest.
- Identify both cross-sectional and longitudinal patterns.
- Try to make the identification of unusual individuals or unusual observations simple.

A number of graphical displays which can be useful in the preliminary assessment of longitudinal data from clinical trials will now be illustrated in the following sections using the four datasets described in the Introduction. But before this we need a small digression to explain how datasets for longitudinal and repeated measures data can be structured in two ways. In the first form there is one observation per subject (typically per person) and the repeated measurements are held in separate variables. We shall refer to this form as the 'wide' form. Alternatively, there may be a separate observation for each measurement occasion, with variables indicating which subject and occasion it belongs to. This is the 'long' form of the dataset. Usually both forms will be needed. The wide form is useful for calculating summary measures, whereas the long form is needed for plots and for fitting appropriate models.

8.3 Glucose Challenge Data

Let's begin with a simple plot that shows the repeated measure profiles of each individual perhaps differentiated in some way according the categories of some grouping variable, for example treatment group; this is known as a *spaghetti plot*. We shall look at several alternative spaghetti plots for the glucose challenge data. In the first we will use different line types for the profiles of the three groups of patients, the relevant code being

```
data pip;
infile "c:\hosgus\data\pip.dat" missover;
  input idno pip1-pip8;
  group=1;
  if idno>13 then group=2;
  if idno>25 then group=3;
run;

data pipl;
  set pip;
  array pips {*} pip1-pip8;
  array t{8} t1-t8 (0.5 1 1.5 2 3 4 5);
  do i=1 to 8;
    time=t{i};
    pip=pips{i};
    output;
  end;
  label time='hours after glucose';
run;

data grplines;
  set pip;
  id='grpline';
  value=idno;
  if group=1 then linepattern='solid';
```

```
  if group=2 then linepattern='dash';
  if group=3 then linepattern='dot';
  keep id value linepattern;
run;

proc sgplot data=pipl dattrmap=grplines noautolegend;
  series y=pip x=time / group=idno attrid=grpline;
  inset "Solid line=controls" "dashed=nonhyperinsulinemic"
    "dotted=hyperinsulinemic";
run;
```

Longitudinal data can be held in two different formats: 'wide' and 'long'. The wide format is where each measurement is held in a different variable, whereas the long format has a separate observation for each measurement occasion. The long format is more useful for graphing longitudinal data although both formats will often be needed.

The glucose challenge data are first read in using the wide format consisting of the person identifier, idno, and eight variables, pip1 to pip8, for the eight measurements of PIP. The second datastep produces a long version of the dataset and adds a time variable for the spacing of the measurements. In this long version of the dataset each person has eight observations.

A basic profile plot, or spaghetti plot, could be produced using the long version of the dataset with proc sgplot as a series plot treating each individual's eight observations as a group and adding noautolegend to the proc sgplot statement. Here we wish to use three line types to distinguish the three treatment groups, but we cannot use the group= option as that is already being used to treat each person's observations as a series. Instead we use an attribute map dataset, grplines. This dataset has one observation per subject but assigns the linepattern based on the three treatment groups. To apply the attribute map to the plot, we specify the dataset, grplines, with the dattrmap= option on the proc sgplot statement and the attribute identifier with the attrid= option on the series plot statement. In more complex examples, the attribute map may contain specifications for more than one attribute and then the id variable would specify which one is to be used. In this case, there is only the one. Rather than attempt to recreate a legend, we put the information in an inset. Each quoted string on the inset statement is put on a separate line.

The resulting plot is shown in Figure 8.1. This version of a spaghetti plot is rather 'messy'. All that can really be concluded is that the PIP values generally decrease over two to three hours and then start to increase again. But it is difficult to assess from Figure 8.1 how the profiles of the three groups of patients compare.

This example also illustrates some of the general problems that commonly beset profile plots. The basic intention is that each individual's profile of measurements can be traced by following a particular line in the plot. Having a large number of

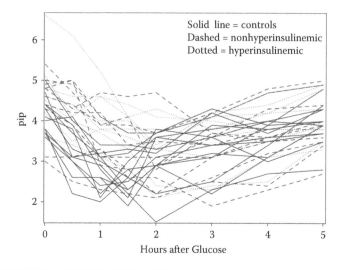

Figure 8.1   Spaghetti plot of glucose challenge data.

subjects in a single plot may make this difficult and in such cases producing separate plots for sub-groups of subjects will usually help. Even taking this to its extreme and plotting each subject individually can be useful in some cases. For example, since there are 33 observations in the glucose challenge dataset, we can fit the individual plots in a 6-by-6 grid as follows:

```
proc sgpanel data=pipl noautolegend;
  panelby idno/rows=6 columns=6 novarname;
  reg y=pip x=time;
  series y=pip x=time;
run;
```

In each panel, we plot the points and fit a simple linear regression line using the reg plot statement and join the points with a **series** plot statement. The resulting plot is shown in Figure 8.2.

Figure 8.2 clearly shows the quadratic nature of the majority of the profiles; only very few are close to the linear fit, for example participant number 28.

A second problem of spaghetti plots is of co-incident points, where two or more profiles pass through the same point (for example, the control group profiles at 2 hours). In such cases, the individual profiles can no longer be distinguished from each other. Separate line types, plotting symbols and/or colours can help but may sometimes result in a plot which is overly 'busy' and confusing to the eye. Jittering—adding a small amount of random error to the *x* and/or *y* values—is another possibility, but the **series** plot statement does not have a jitter option so the jittered values would need to be calculated separately.

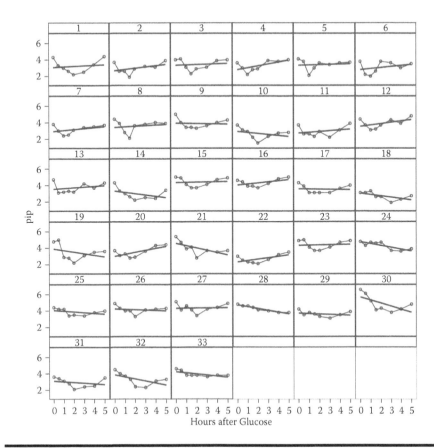

Figure 8.2   Individual participant profiles for the glucose challenge data.

For the glucose challenge data an obvious alternative to Figure 8.1 is to plot the profiles of each patient group on different panels of the same plot. We use **proc sgpanel** to achieve this as follows:

```
proc format;
  value grp 1='Control'
          2='nonhyperinsulinemic'
          3='hyperinsulinemic';
run;

ods graphics/height=7.5cm width=16cm;
proc sgpanel data=pipl noautolegend;
  panelby group/rows=1 novarname;
  series y=pip x=time/group=idno;
  format group grp.;
run;
ods graphics/reset;
```

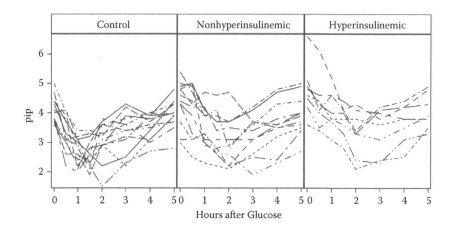

Figure 8.3   Individual participant profiles for the glucose challenge data by group.

First we create a format to label the three patient groups. The ods graphics statement sets the overall height and width of the panel plot so that the individual plots are less stretched. The panelby statement names group as the variable to form the panels within the plot. The panels are to be arranged in a single row so that it is easier to compare the *y*-axis values. In this case we have chosen to allow different line types for each subject, but we could have kept them all solid lines by specifying this with the lineattrs= option. Finally, the graphic options are reset to their defaults.

The resulting plot is shown in Figure 8.3. The plots again show the quadratic nature of the participants' profiles and suggest that the profiles in the control group have a somewhat different shape than those in the other two groups. Such information is very valuable to have before proceeding to fit formal models to the data.

As well as producing separate plots for existing groups of individuals, it may be informative to subdivide by some feature of the response. We can illustrate this with a plot where the profiles of individuals with 'low', 'medium' and 'high' baseline values (simply defined in terms of an equal division of baseline values into three) are plotted separately. Such a plot is constructed using the following code:

```
proc rank data=pipl out=pipl groups=3;
  var pip1;
  ranks tertile;
run;
proc format;
  value tertile 0='low' 1='medium' 2='high';
run;
proc sgpanel data=pipl noautolegend;
  panelby tertile group/layout=lattice;
  series y=pip x=time/group=idno lineattrs=(pattern=solid);
```

```
format tertile tertile.;
 label tertile='Baseline';
run;
```

Although we are working with the 'long' version of the dataset, it still contains all eight measurements within each observation, including the initial measurement, pip1, so that we can subdivide the observations into tertiles using **proc rank** with the **groups=3** option. The **ranks** statement stores the values in the variable, **tertile**.

The plot in Figure 8.4 suggests that participants with high baseline values tend to decrease somewhat over time making the final PIP value less than the baseline value. Patients with low and medium baseline values tend to increase or stay the same over time. This may simply reflect the *regression to the mean* phenomena (see Everitt and Skrondal, 2010).

With large numbers of observations, graphical displays of individual response profiles soon become of little use except for a general appraisal of the variation in the data; so plots of average group profiles along with some indication of the

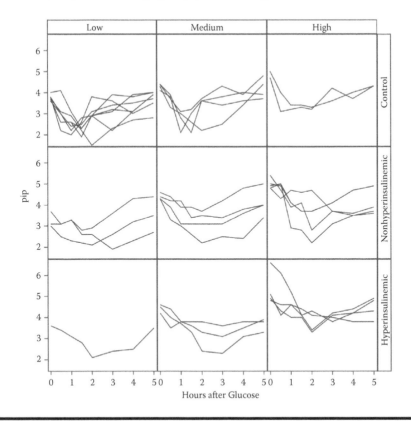

Figure 8.4   **Profiles plots for the glucose challenge data classified by patient group and size of baseline value.**

variation of the observations at each time point become essential for judging how the groups of observations change over time and for indicating any differences between the groups. Such a graph can be constructed using the following code:

```
proc sgplot data=pipl;
vline time/response=pip group=group stat=mean limits=both limitstat=stderr
  numstd=1.4 groupdisplay=cluster clusterwidth=.4;
  format group grp.;
  xaxis type=linear;
run;
```

The vline statement within **proc sgplot** can be used to plot summary statistics. The *x*-axis variable follows the vline keyword and is assumed to be categorical. The remainder of the plot uses two sets of options. The first set specifies the summary values: response=pip names pip as the variable to summarize on the *y*-axis; stat=mean selects the mean as the summary statistic; limitstat=stderr adds standard error limit lines; and numstd=1.4 makes these 1.4 standard errors above and below the mean. The second set of options controls the display of groups: the grouping variable is called group; groupdisplay=cluster offsets the groups on the *x*-axis and clusterwidth=.4 reduces the amount of space between groups from the default of .8.

We have chosen 1.4 standard errors following Cumming (2009) as this gives a better visual indication of when the groups differ significantly (at the 5% level) than the more usual 95% confidence limits do. Offsetting the groups slightly on the *x*-axis is useful when there are varying degrees of overlap but runs the risk of distorting the information on timing if the clusters are too wide. The xaxis statement with type=linear ensures that the time points are correctly spaced, whereas the default treatment of the *x*-axis as categorical would have made them equally spaced.

The resulting plot is shown in Figure 8.5. We see that the two patient groups reach, on average, their lowest PIP value about 30 minutes later than the control group and all three groups recover, approximately, to their initial PIP values after 5 hours. The variation in the PIP values in each group appears to be very similar. The control group drops significantly lower than the other two groups following the challenge then reverts to similar values at 2 hours.

A possible alternative to Figure 8.5 is to plot side-by-side boxplots for each group at each time a measurement is made. Such a plot is constructed with the following code:

```
proc sgpanel data=pipl;
  panelby group/columns=3 novarname;
  vbox pip/category=time nocaps nomean;
  colaxis type=linear;
  format group grp.;
run;
```

The resulting plot is shown in Figure 8.6. This plot gives essentially the same message as Figure 8.5, namely the quadratic nature of the profiles, reaching the

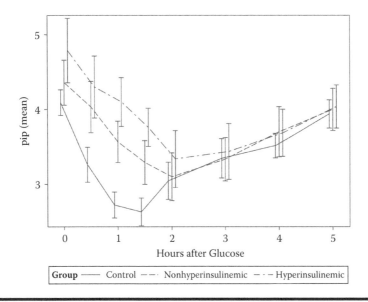

Figure 8.5   Average PIP profiles of the three groups in the glucose challenge data showing the variation in the PIP values.

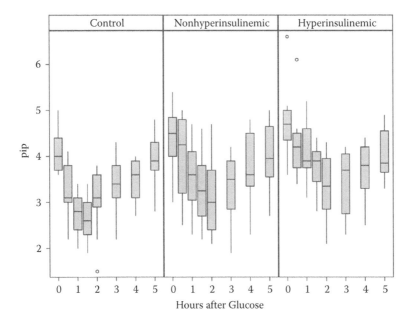

Figure 8.6   Boxplots for each occasion of the glucose challenge data and for each group of participants.

minimum PIP value earlier in the control group and the return to the baseline value after 5 hours.

The two graphs produced for the glucose challenge data suggest that models for the data should contain quadratic time effects and possibly Group × Time interactions. And as described in Der and Everitt (2013) such models need to adequately account for the pattern of correlations between the repeated measurements. Some idea of what will be needed can be obtained by producing the scatterplot matrices of the repeated measures, separately for each group. This can be done as follows:

```
proc sgscatter data=pip;
  matrix pip1-pip8;
  by group;
run;
```

The resulting plots are shown in Figure 8.7.

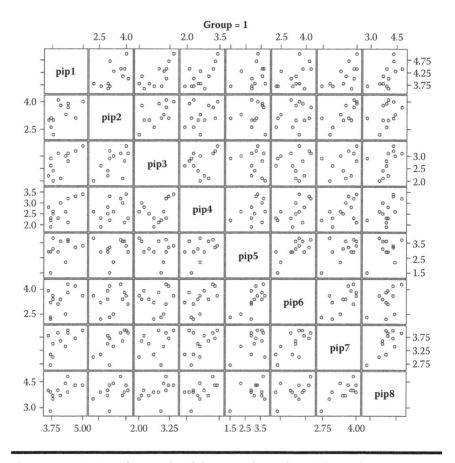

Figure 8.7a Scatterplot matrix of the PIP values of participants in group one in the glucose challenge data.

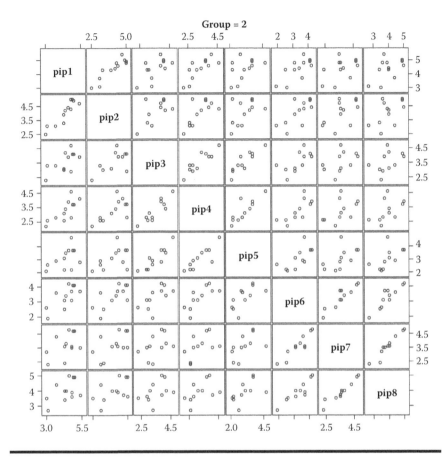

Figure 8.7b (*Continued*) **Scatterplot matrix of the PIP values of participants in group two in the glucose challenge data.**

The various scatterplot matrices show that adjacent PIP values are very highly related and that the relationship between non-adjacent time points is generally not as strong although the association between the baseline value and the final value is relatively high in all three groups. Clearly fitting a model that assumed the repeated measurements were independent of one another would be inappropriate here and it also appears that a model assuming that each pair of repeated measures has *equal* correlation (*compound symmetry*) is not suitable.

8.4 CBT for Depression

For the CBT for depression data we will begin with a spaghetti plot, but one designed to reveal a common feature of longitudinal data, namely the ability to predict subsequent observations from earlier ones. Informally this implies that

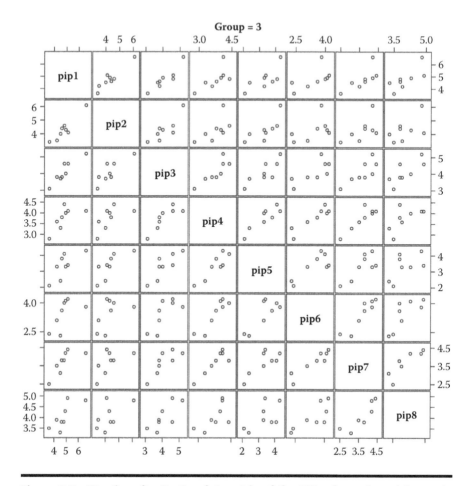

Figure 8.7c (*Continued*) Scatterplot matrix of the PIP values of participants in group three in the glucose challenge data.

individuals that have, for example, the largest values of the response variable at the start of the study to continue to have the larger values throughout the study. The phenomenon is generally referred to as *tracking*. (In more formal terms, a population is said to track with respect to a particular observable characteristic if, for each individual, the expected value of the relevant deviation from the population mean remains unchanged over time.) Identification of tracking in a set of longitudinal data can often be helpful in constructing appropriate models for the data.

The tracking feature of longitudinal data can be seen more clearly if we plot the *standardized* values of each observation, i.e. the values obtained by subtracting the relevant visit mean from the original observation and then dividing by the corresponding visit standard deviation. The following code produces such a plot for the CBT data:

```
data btb;
  infile "c:\hosgus\data\btb.dat" expandtabs missover;
  input idno Treatment$ BDIpre BDI2m BDI3m BDI5m BDI8m;
run;

data btbl;
  set btb;
  array bdis {*} BDIpre -- BDI8m;
  array t {*} t1-t5 (0 2 3 5 8);
  do i=1 to 5;
    bdi=bdis{i};
    time=t{i};
    if i<5 and bdis{i+1}~=. then next=1;
    else next=0;
    output;
  end;
  drop BDI2m -- BDI8m t1-t5 i;
run;

proc sort data=btbl;
  by time;
run;
proc stdize data=btbl out=btblz;
  var bdi;
  by time;
run;
proc sort data=btblz;
  by idno time;
run;
proc sgpanel data=btblz noautolegend;
  panelby treatment/rows=1 spacing=10 novarname;
  series y=bdi x=time/group=idno lineattrs=(pattern=solid);
run;
```

We begin by reading in the data, which are initially in the wide format. The second datastep produces the long format as before with the PIP dataset. An additional variable, **next**, is created, which indicates whether or not the subject attended the following visit. This will be used later. The dataset is then sorted by time, as we wish to standardize it separately within each time point with **proc stdize**. The dataset with the standardized values, **btblz**, is then sorted back to the original order and **proc sgpanel** used for the spaghetti plot. The **lineattrs** option results in a solid line being used for all the profiles.

The resulting plot is shown in Figure 8.8.

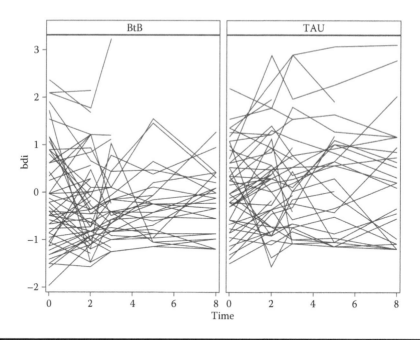

Figure 8.8 Spaghetti plot of individual profiles of participants in the CBT data after standardization.

The tracking feature of the data is clearly demonstrated in Figure 8.8—the majority of the standardized profiles are approximately horizontal.

Next we produce a plot of the average BDI values for the two treatment groups over time using the code:

```
proc sgplot data=btbl;
  vline time/response=bdi group=treatment stat=mean limitstat=stderr
  numstd=1.4;
  xaxis type=linear;
run;
```

The resulting plot is shown in Figure 8.9. (This has error bars of 1.4 SE, again following Cumming, 2009.)

Figure 8.9 suggests that the BDI values in both treatment groups decline over time but this decline is greater in the BtB group than in the TAU group.

An important feature of the BtB data that we now need to consider is the presence of missing values caused by subjects dropping out of the study. We can plot the proportion missing in each treatment group over time using the code:

```
data btbl;
  set btbl;
  bdim=missing(bdi);
run;
```

```
proc sgplot data=btbl;
  vline time/response=bdim group=treatment stat=mean;
  xaxis type=linear;
  yaxis label='Proportion missing';
run;
```

The result is shown in Figure 8.10. Initially more people drop out of the BtB treatment group but later the proportion of people dropping out is much the same in each group.

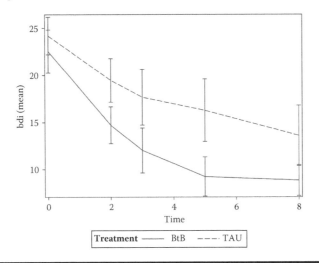

Figure 8.9   **Average BDI values over time for the two treatments in the CBT data.**

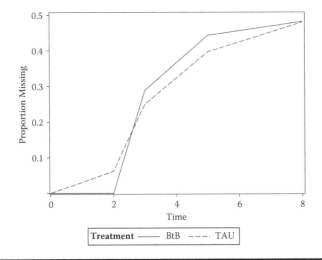

Figure 8.10   **Proportion of dropouts in the two treatment groups in the CBT study.**

The reason why values are missing, in particular whether these reasons relate to recorded (non-missing) values, has implications for which potential models are suitable and which are not. Details of the different types of missing data mechanism are given in Der and Everitt (2013). Here we consider just one, *missing at random*, in which the missing values are independent of the observed or unobserved values; in other words the probability of being missing is the same for all members of the sample. All models that might be considered for longitudinal data are valid under the missing at random assumption so it becomes of importance to try to see if a dataset with such values meets the assumption. Carpenter, Pocock and Lamm (2002) suggest a relatively simple plot for assessing whether dropout is not completely at random. Values of the response variable for the subjects in each treatment group are plotted at each time point (including pre-randomisation), differentiating on the plot those who do and those who do not attend their next scheduled visit; any clear difference in the distribution of values for 'attenders' and 'nonattenders' would indicate that 'missingness' in the data is not completely random. We can construct this plot for the BtB data as follows:

```
proc format;
  value yn10f 0='No' 1='Yes';
run;
proc sgpanel data=btbl;
  panelby treatment/rows=2 spacing=10 novarname;
  scatter y=bdi x=time/group=next groupdisplay=cluster
    grouporder=descending clusterwidth=.4;
  label next='Attended next visit';
  format next yn10f.;
  where time<8;
run;
```

The variable **next** is used as a **group** variable to distinguish those who did and did not attend the following visit. The **groupdisplay=cluster** option offsets them on the *x*-axis and the **clusterwidth** is reduced from its default of .8 to bring the groups closer together. The final time point is not applicable to this comparison and so is excluded from the graph. A format and labels are used to make the graph self-explanatory. The result is shown in Figure 8.11.

In Figure 8.11 we can compare the distribution of BDI values for patients who do and those who do not attend their next scheduled visit and there is no apparent difference. Consequently it is probably reasonable to assume dropout is completely random, which has implications for which types of analysis are appropriate for these data.

8.5 Graphing Residuals from Fitting Models to Longitudinal Data

After the initial assessment of a longitudinal dataset using the graphics described in the previous sections, a researcher will need to move on to a more formal

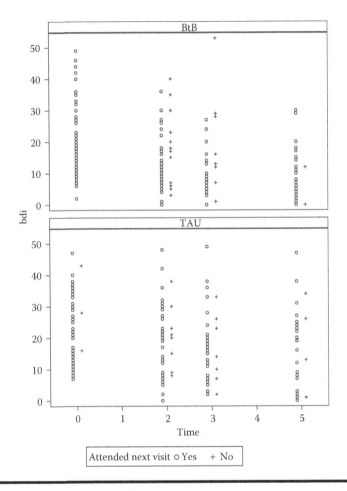

Figure 8.11 Distribution of BDS values for patients who do and do not attend their next scheduled visit in the CBT data.

analysis of the data using suitable models often suggested, in part at least, by the graphic displays constructed for the data. An account of such models and how they are fitted using SAS is given in Der and Everitt (2013). But as with fitting multiple regression models (see Chapter 6) the analysis of longitudinal data is not complete without an examination of residuals. The situation is, however, more complicated than that described in Chapter 6 because now associated with each individual there is a *vector* of residuals, r_i say for the ith individual, the elements of which are the differences between the observed and fitted values of the response variable at each recorded time point. (If the number of repeated measurements is the same for each individual, m say, then each vector of residuals will have m elements.)

Graphical displays of these residuals can be used to check for any systematic departures from the model for the mean response or to detect the presence of outlying *observations* that need further investigation and also to possibly detect outlying *individuals*.

To illustrate the use of residuals in assessing and diagnosing models fitted to longitudinal data we shall use the glucose challenge data given in Table 8.1 but simplified a little by amalgamating the 12 non-hyperinsulinemic obese patients and 8 hyperinsulinemic patients into a single 'obese' group. To begin we will plot the patient profiles of the members of the two groups using the code

```
data pipl;
  set pipl;
  group=1;
if idno>13 then group=2;
run;
proc format;
  value twogrp 1='Control' 2='Obese';
run;
proc sgpanel data=pipl noautolegend;
  panelby group/rows=1 spacing=10 novarname;
  series y=pip x=time/group=idno lineattrs=(pattern=solid);
  format group twogrp.;
run;
```

The resulting plot is shown in Figure 8.12.

The quadratic nature of the profiles is clear in this plot (as it was in earlier graphics for these data) so clearly a quadratic effect is needed in any sensible model, and in Der and Everitt (2013) it is shown that a Group × Time interaction is needed. So the model we shall fit to the data is specifically

$$y_{ij} = (\beta_0 + u_{i1}) + \beta_1 \text{group} + (\beta_2 + u_{i1})\ \text{time} + \beta_3 \text{time}^2 + \beta_4 (\text{group} \times \text{time}) + \varepsilon_{ij} \quad (8.1)$$

which is a mixed effects model with random slope and intercept, linear and quadratic time effects, and a Group × Time interaction where 'group' is a dummy variable that identifies whether a subject is in the control or obese group.

A detailed account of fitting mixed effects models using SAS is given in Der and Everitt (2013) and the model in Equation 8.1 can be fitted to the glucose challenge data using the following SAS code:

```
proc mixed data=pipl covtest noclprint;
  class group idno;
  model pip=group|time|time@2/s ddfm=bw residual outp=mixout;
  random int time/subject=idno type=un;
run;
```

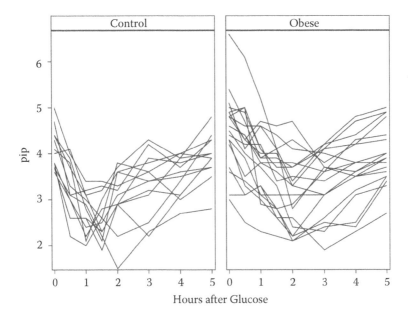

Figure 8.12 Individual profiles for glucose challenge data after combining the two patient groups into a single obese group.

The numerical results are given in Der and Everitt (2013) and the interaction effect is highly significant. Here, however, we will concentrate on various graphical displays for assessing whether or not the model is a reasonable one to fit to the data. To begin we can plot the fitted values using the code:

```
proc sgpanel data=mixout noautolegend;
  panelby group/rows=1 spacing=10 novarname;
  series y=pred x=time/group=id lineattrs=(pattern=solid);
  format group twogrp.;
run;
```

The resulting plot is shown in Figure 8.13. Clearly the predicted profiles reflect quite well the observed profiles given in Figure 8.12 with the predicted profiles of the obese group being somewhat 'flatter' than those in the control group.

Before looking at any residual plots for this model we will consider the two random effects in the model; both of these are assumed to be normally distributed. But the random effects are not estimated as part of the model. However, having fitted the model, we can predict the values of the random effects by predicting its conditional mean given the available data; full details are given in Fitzmaurice et al. (2004).

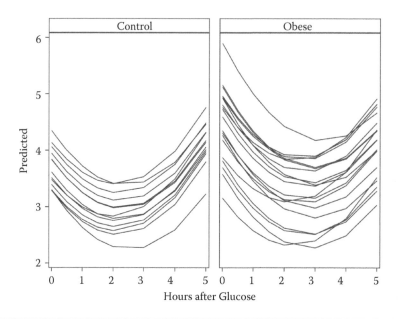

Figure 8.13   Predicted profiles from fitting model in Equation 8.1.

Using the predicted random effect values we can construct probability plots to assess their normality using the following code:

```
proc mixed data=pipl covtest noclprint;
  class group idno;
  model pip=group I time I time@2/s ddfm=bw residual outp=mixout;
  random int time/subject=idno type=un s;
ods output solutionr=reffs;
run;

proc univariate data=reffs(rename=(estimate=Intercept) where=
  (Effect='Intercept'))noprint;
  probplot Intercept/normal(mu=est sigma=est);
run;
proc univariate data=reffs(rename=(estimate=Slope) where=(Effect='time'))
  noprint;
  probplot Slope/normal(mu=est sigma=est);
run;
```

The dataset **reffs** contains the predicted random effect values for both the intercept and the slope in **time**. Proc **univariate** is used to produce the normal probability plots with the **rename** and **where** dataset options used to select one or other, and rename the **estimate** variable accordingly.

The plots are shown in Figure 8.14.

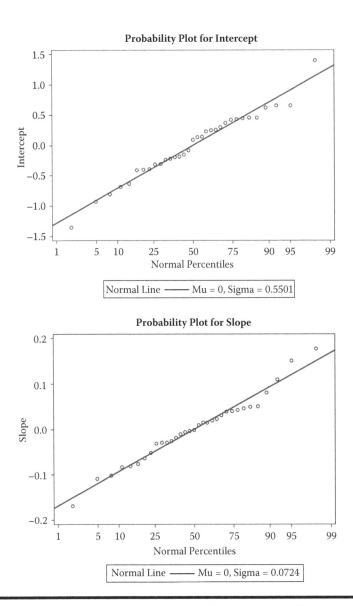

Figure 8.14 Normal probability plots for predicted random effects for intercept and slope in the glucose challenge data.

The plots of the residuals are each essentially linear as required, although there is some *slight* deviation from linearity in each plot.

A further plot that can be helpful is a scatterplot of the residuals against predicted values. In a correctly specified model the scatterplot should not display any systematic pattern; the fitting of a loess curve (see Chapter 6) can often help in assessing the scatterplot. Similarly scatterplots of the residuals against selected covariates from the model for the mean response can be examined for systematic trends, which, if present, may indicate the omission of a quadratic term or the need for transformation of a covariate. We can construct the residual against predicted and residual against time scatterplots for the final model fitted to the glucose challenge data using the code

```
proc sgpanel data=mixout noautolegend;
  panelby group/rows=1 spacing=10 novarname;
  loess y=resid x=pred;
  format group twogrp.;
run;
```

```
proc sgpanel data=mixout noautolegend;
  panelby group/rows=1 spacing=10 novarname;
  loess y=resid x=time;
  format group twogrp.;
run;
```

The plots are shown in Figures 8.15.

There are no clear patterns in either plot that may cause concern about the validity of the fitted model.

To show how the residual plots can give an indication of a misspecified model we will now consider the corresponding plots for a model that includes the Group × Time interaction, but only a linear effect for time. The code is very similar to that used earlier and produces Figure 8.16.

Here the plot of residuals against time shows a clear pattern indicating that a quadratic effect is indeed needed in the model.

Finally we can construct plots for the model predicted individual profiles from the model in Equation 8.1 and the corresponding model without the quadratic time effect.

```
data bothmod;
  set mixout mixout2(in=in2);
  model='Quadratic';
  if in2 then model='Linear';
run;
```

```
proc sgpanel data=bothmod noautolegend;
  panelby model group/rows=2 layout=lattice;
  series y=pred x=time/group=idno;
  format group twogrp.;
run;
```

A short datastep combines the predicted values from the two models and assigns a variable to indicate which model provided the values. This is used together with group to form the panels. A lattice layout is specified so that the values of model form the columns and group the rows. The result is shown in Figure 8.17.

With a small dataset such as this, the predicted profiles and residuals can be displayed individually in a panel plot in a similar fashion to Figure 8.2.

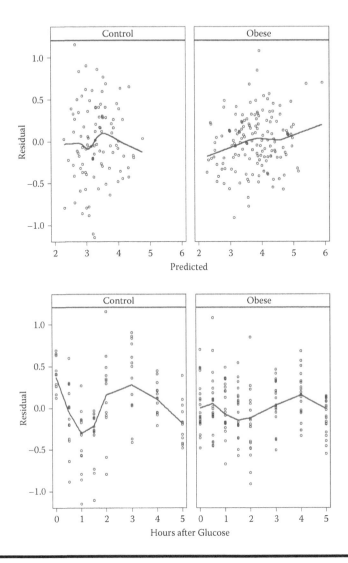

Figure 8.15 Various residual plots for glucose challenge data.

```
proc sgpanel data=mixout noautolegend;
  panelby idno/rows=6 columns=6 novarname;
  band x=time upper=upper lower=lower;
  series y=pred x=time;
  scatter y=pip x=time;
run;
```

The confidence band needs to be plotted before the predicted and observed values so that it does not overwrite them. The result is shown in Figure 8.18.

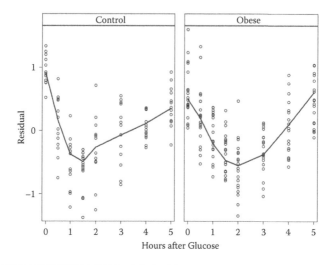

Figure 8.16 Residual plots for the glucose challenge data under a model that allows only a linear effect of time.

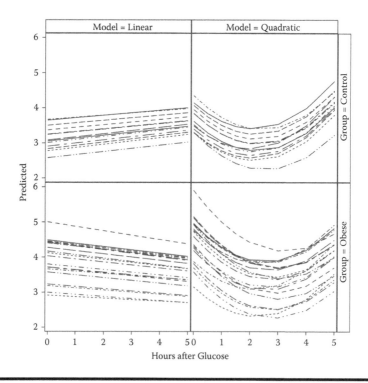

Figure 8.17 Predicted profiles from the linear and quadratic models.

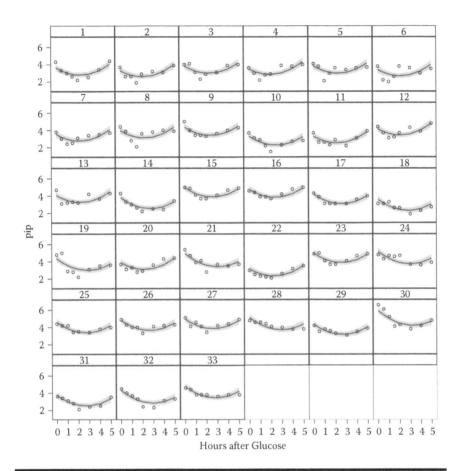

Figure 8.18   **Observed and predicted values, with 95% confidence bands, for the quadratic model fitted to the glucose challenge data.**

8.6 Summary

Analysing longitudinal data can often be a tricky prospect particularly when the data contain missing values. The graphs described in this chapter are essential aids in both understanding the data and suggesting what might be sensible models to consider and in interpreting the numerical results from such models.

Exercises

8.1. The data in Table 8.3 are a sub-set of the data collected in a clinical trial of the use of oestrogen patches in the treatment of post-natal depression; full details of the study are given in Gregoire et al. (1996). Use any of the graphics described in the chapter to gain insight into the message in the data and also use them to indicate which models you think may be suitable for the data. Do you think the treatment works?

Table 8.3 Data from a Clinical Trial of Oestrogen Patches in the Treatment of Post-Natal Depression

Group	EPDS score							
0 = placebo				Month				
1 = treatment	Pre1	Pre2	1	2	3	4	5	6
0	18.00	18.00	17.00	18.00	15.00	17.00	14.00	15.00
0	25.00	27.00	26.00	23.00	18.00	17.00	12.00	10.00
0	19.00	16.00	17.00	14.00	–9.00	–9.00	–9.00	–9.00
0	16.00	18.00	19.00	–9.00	–9.00	–9.00	–9.00	–9.00
....								
1	21.00	21.00	13.00	12.00	9.00	9.00	13.00	6.00
1	27.00	27.00	8.00	17.00	15.00	7.00	5.00	7.00
1	24.00	15.00	8.00	12.00	10.00	10.00	6.00	5.00
....								

8.2. Five different types of electrodes were applied to the arms of 16 subjects and the resistance (in kilohms) measured. The data are given in Table 8.4. The experiment was designed to see whether the different types of electrodes performed similarly. Explore the data graphically using graphical displays to answer (informally) the question of interest. Look out for possible outliers!

Table 8.4 Skin Resistance Data

1	500	400	98	200	250
2	660	600	600	75	310
3	250	370	220	250	220
4	72	140	240	33	54
5	135	300	450	430	70
6	27	84	135	190	180

Table 8.4 (*Continued*)   Skin Resistance Data

7	100	50	82	73	78
8	105	180	32	58	32
9	90	180	220	34	64
10	200	290	320	280	135
11	15	45	75	88	80
12	160	200	300	300	220
13	250	400	50	50	92
14	170	310	230	20	150
15	66	1000	1050	280	220
16	107	48	26	45	51

Chapter 9

Graphs for Survival Data: Motion Sickness and Breast Cancer

9.1 Introduction

In this chapter we will look at two examples of *survival data*. Such data involve times to the occurrence of some event, frequently death but not always, as we will see in the first dataset. Survival data are characterised by generally having a distribution that is positively skewed and by containing *censored observations*, i.e. individuals who at the end of the study are still 'alive' so that all can be said about the survival time is that it is longer than the recorded time.

Motion sickness—Burns (1984) reported on experiments performed as part of a research programme investigating motion sickness at sea. Subjects were placed in a cubical cabin mounted on a hydraulic piston and subjected to vertical motion for two hours. The length of time until each subject first vomited was recorded. Censoring occurred because some subjects requested an early end to the experiment, while others survived the whole two hours without vomiting. (These data demonstrate that the end point of 'survival data' is not necessarily death.) The data shown in Table 9.1 come from two experiments, one with 21 subjects experiencing motion at a frequency of 0.167 Hz and acceleration 0.111 g and the other with 28 subjects moving at frequency 0.333 Hz and acceleration 0.222 g. Censored observations are indicated by an asterisk (*). Interest lies in how the different types of motion affect time to vomiting.

Table 9.1 Time to Vomiting in an Investigation of Motion Sickness

Experiment					
1			*2*		
1	30		1	5	
2	50		2	6	*
3	50	*	3	11	
4	51		4	11	
5	66	*	5	13	
6	82		6	24	
7	92		7	63	
8	120	*	8	65	
9	120	*	9	69	
10	120	*	10	69	
11	120	*	11	79	
12	120	*	12	82	
13	120	*	13	82	
14	120	*	14	102	
15	120	*	15	115	
16	120	*	16	120	*
17	120	*	17	120	*
18	120	*	18	120	*
19	120	*	19	120	*
20	120	*	20	120	*
21	120	*	21	120	*
			22	120	*
			23	120	*
			24	120	*
			25	120	*
			26	120	*
			27	120	*
			28	120	*

*Censored observations.

Breast cancer—The data here arise from a randomized clinical trial investigating the effects of hormonal treatment with Tamoxifen in women suffering from node-positive breast cancer (Schumacher et al., 1994). Data from randomised patients from this trial and additional non-randomised patients from the German Breast Cancer Study Group 2 (GBSG2) make up the 686 women in the dataset. As well as the survival time of each patient the values of seven other variables (*covariates*) were also recorded for each of the women in the study (Sauerbrei and Royston, 1999). These variables are age at the start of the study, menopausal status, tumour size, tumour grade, number of positive lymph nodes, progesterone receptor, oestrogen receptor and whether or not the patient received hormonal therapy. A small subset of the data is given in Table 9.2.

Table 9.2 Subset of the German Breast Cancer Data

ID	Horm	Age	Menstat	Tsize	Tgrade	Pnodes	Progrec	Estrec	Time	Status
1	0	70	Post	21	II	3	48	66	1814	1
2	1	56	Post	12	II	7	61	77	2018	1
3	1	58	Post	35	II	9	52	271	712	1
4	1	59	Post	17	II	4	60	29	1807	1
5	0	73	Post	35	II	1	26	65	772	1
6	0	32	Pre	57	III	24	0	13	448	1
7	1	59	Post	8	II	2	181	0	2172	0
8	0	65	Post	16	II	1	192	25	2161	0
9	0	80	Post	39	II	30	0	59	471	1
10	0	66	Post	18	II	7	0	3	2014	0
11	1	68	Post	40	II	9	16	20	577	1
12	1	71	Post	21	II	9	0	0	184	1
13	1	59	Post	58	II	1	154	101	1840	0
14	0	50	Post	27	III	1	16	12	1842	0
15	1	70	Post	22	II	3	113	139	1821	0
16	0	54	Post	30	II	1	135	6	1371	1
17	0	39	Pre	35	I	4	79	28	707	1
18	1	66	Post	23	II	1	112	225	1743	0
19	1	69	Post	25	I	1	131	196	1781	0
20	0	55	Post	65	I	4	312	76	865	1

Note: Horm, dichotomous variable indicating whether hormonal therapy was applied, 0 = no, 1 = yes; Age, age in years; Menstat, menopausal status, post or pre; tsize, tumour size; Tgrade, tumour grade; Pnodes, number of positive lymph nodes; Progrec, progesterone receptor; Estrec, estrogen receptor; Time, survival time in days; Status, whether alive (0) or dead (1) at end of study.

9.2 The Survival Function

The first step in the analysis of a set of survival times is almost always the construction of a graphic used to describe the distribution of survival times, a graphic which essentially estimates what is known as the *survivor* (or *survival*) *function* of the data. The population survivor function, $S(t)$, is defined as the probability that the survival time, T, is greater than or equal to t, i.e.

$$S(t) = \Pr(T > t) \tag{9.1}$$

Before we move on to how this function is estimated from a sample of survival times it will be helpful to see what it looks like for two probability distributions often used to model survival time data, namely the *exponential distribution* and the *Weibull distribution*. First, the exponential with probability density function

$$f(t) = \lambda e^{-\lambda t}, \quad 0 \le t < \infty \tag{9.2}$$

for which the survivor function is given by

$$S(t) = \int_t^\infty \lambda e^{-\lambda u} du = e^{-\lambda t} \tag{9.3}$$

Plots of the survivor functions of the exponential distribution for different values of λ are shown in Figure 9.1.

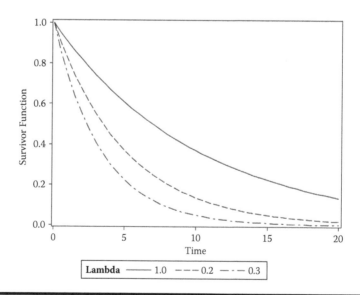

Figure 9.1 Survival functions for a number of exponential distributions.

Now for the Weibull probability density function, which is given by

$$f(t) = \lambda\gamma t^{\gamma-1} \exp(-\lambda t^{\gamma}), \quad 0 \leq t < \infty \tag{9.4}$$

The survivor function of the Weibull is given by

$$S(t) = \int_t^{\infty} \lambda\gamma u^{\gamma-1} \exp(-\lambda u^{\gamma}) du = \exp(-\lambda t^{\gamma}) \tag{9.5}$$

Plots of the survivor function of the Weibull distribution for different values of the scale parameter, λ, and the shape parameter, γ, are shown in Figure 9.2.

The Weibull distribution can clearly represent the survival distributions of a wider range of populations of survival times than the exponential making it potentially a more useful model for such datasets.

When we have a sample of survival times a plot of an estimate of $S(t)$ against t is often a useful way of describing the survival experience of the individuals in the sample and for comparing the survival experience of different groups of subjects of interest, for example men and women. When there are no censored observations in a sample of n survival times, a *nonparametric estimate* (i.e. does not require specific assumptions about the distribution of the survival times) of the survivor function is given by

$$\hat{S}(t) = \frac{\text{number of individuals with survival times} \geq t}{\text{number of individuals in the dataset}} \tag{9.6}$$

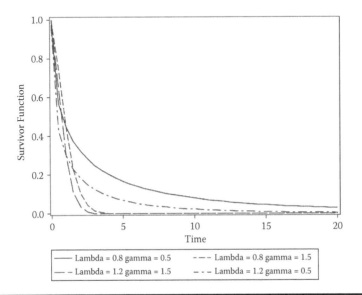

Figure 9.2   Survival functions for a number of Weibull distributions.

with the convention that $\hat{S}(t) = 1$ for t less than the smallest observed survival time. Because this is simply a proportion, confidence intervals can be obtained for each time t by using the variance estimate

$$\hat{S}(t)\left(1-\hat{S}(t)\right)/n \qquad (9.7)$$

The estimated survivor function, $\hat{S}(t)$, is assumed to be constant between two adjacent death times with the consequence that a plot of t against $\hat{S}(t)$ is a *step-function*, which decreases immediately after each observed survival time.

The aforementioned simple method cannot be used to estimate the survivor function when the data contains censored observations and it is such observations that are the quintessential feature of survival data. In the presence of censoring, the survivor function is generally estimated using the *Kaplan–Meier estimator* (also known as the *product-limit estimator*), which again is nonparametric. This estimator is based on the calculation and use of conditional probabilities and incorporates information from all the observations available, both censored and uncensored, by considering survival to any point in time as a series of 'steps', which are intervals defined by a rank ordering of the survival times. So we denote by $t_1 < t_2$K, the times when 'deaths' occurred, and let d_j be the number of individuals who die at time t_j and then the Kaplan–Meier estimator for the survivor function is given by

$$\hat{S}(t) = \prod_{t_j \le t}\left(1-\frac{d_j}{r_j}\right) \qquad (9.8)$$

where r_j is the number of individuals at risk, i.e. alive and not censored, just prior to time t_j. If there are no censored observations, the estimator in Equation 9.8 reduces to that in Equation 9.6. The estimated variance of the Kaplan–Meier estimator is given by

$$V\left[\hat{S}(t)\right] = [\hat{S}(t)]^2 \sum_{t_j \le t}\frac{d_j}{r_j(r_j - d_j)} \qquad (9.9)$$

When there is no censoring, this reduces to the variance estimator given in Equation 9.7. A $100(1 - \alpha)\%$ confidence interval for $S(t)$, for a given value of t is given by the interval $\hat{S}(t) \pm z_{\alpha/2}\sqrt{V\left[\hat{S}(t)\right]}$, where $z_{\alpha/2}$ is the upper $\alpha/2$ value of the standard normal distribution. These intervals can be superimposed on a graph of the estimated survivor function as we shall see later. Collett (2004) points out a potential problem with this procedure that arises from the fact that the confidence intervals are symmetric and when the survivor function is close to zero or unity symmetric intervals are inappropriate because they can lead to confidence intervals for the survivor function that lie outside the interval (0,1). Collett offers as a pragmatic solution, replacing any limit that is greater than unity by 1.0 and any limit that is less than zero by 0.0. Collett also describes some alternative approaches to constructing confidence intervals for the survivor function.

9.3 The Survival Function of the Motion Sickness Data

We can construct and plot the estimated survivor functions of the two groups of subjects in the motion sickness data using the following code:

```
data motion;
   infile "c: \ hosgus \ data \ motion.dat" missover;
   input idno minutes star $;
   censor=0;
   if star='*' then
      censor=1;
   if _n_<22 then experiment=1;
      else experiment=2;
run;

proc lifetest data=motion plots=survival;
   time minutes*censor(1);
   strata experiment;
run;
```

The data in the file **motion.dat** are stacked in order of the experiments and are read in as three columns.

Proc lifetest is used to estimate and plot the survivor function as well as for testing differences in survival between groups. A range of **plots** is available with the plots option, including a smoothed estimate of the hazard function described later. Here we simply request the survivor function and this could be abbreviated to **plots=s**. The **time** statement is used to specify survival time and censoring. The variable containing the survival times comes first, then an asterisk and the censoring variable with a value, or list of values, indicating censored observations in parentheses. The censoring variable needs to be numeric and have valid values (i.e. not missing) for both censored and non-censored observations. The different groups in the data for which separate survival curves are to be plotted are specified on the **strata** statement.

The resulting plot is shown in Figure 9.3.

Censored observations are indicated by a plus (+) on the plot. Figure 9.3 shows that 'survival time' (time to vomiting in this example) appears to be longer in the first group of subjects, those subjected to motion at a frequency of 0.167 Hz and acceleration 0.111 g. But without looking at some measure of the variation in the estimated survival functions, assessing the difference between them is largely subjective. A more objective assessment of the difference in 'survival' of the two groups

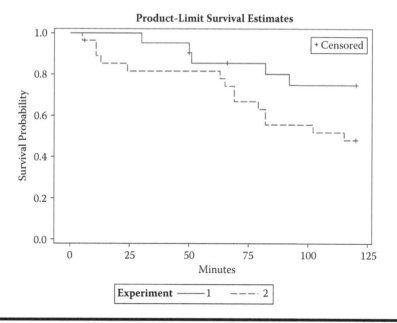

Figure 9.3   Estimated survival functions for each group of subjects in the motion sickness example.

can be made if we add the confidence intervals for each curve to Figure 9.3 and this can be done with the **cl** option. This example also illustrates the **test** option to display the test result on the plot. The new **proc** statement is

```
proc lifetest data=motion plots=s(cl test);
```

The result is shown in Figure 9.4. Now the apparent difference between the two survival functions suggested by Figure 9.3 is thrown into doubt because of the large overlap of the two confidence intervals and it appears that there is no real difference in the time to vomiting produced by the two types of motion inflicted on the subjects in the two groups. A formal test of the equality of the survival functions using the *log-rank test* (see Der and Everitt, 2009) shows that the data are consistent with the hypothesis of equality of survival functions, as the associated *p*-value of the test (shown on Figure 9.4) is 0.0733.

9.4 The Hazard Function

In the analysis of survival data it is often of interest to assess which periods have high or low chances of death (or whatever the event of interest may be) among those still active at the time. A suitable approach to characterize such risks is the *hazard*

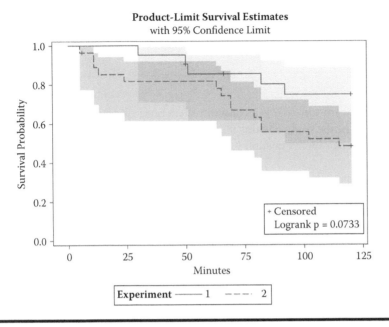

Figure 9.4   Estimated survival curves and confidence intervals for the two groups of subjects in the motion sickness example.

function, h(*t*), defined as the probability that an individual experiences the event in a small time interval *s*, given that the individual has survived up to the beginning of the interval, when the size of the time interval approaches zero. Mathematically this is written as

$$h(t) = \lim_{s \to 0} \frac{\Pr\left(t \le T \le t + s \mid T \ge t\right)}{S} \tag{9.10}$$

where *T* is the individual's survival time. The conditioning feature of this definition is very important. For example, the probability of dying at age 100 is very small because most people die before that age; in contrast, the probability of a person dying at age 100 who has reached that age is much greater.

The hazard function is a measure of how likely an individual is to experience an event as a function of the age of the individual; it is often known as the *instantaneous death rate.*

Collett (2004) shows that the hazard function can be given in terms of a probability density function and the corresponding survivor function as

$$h(t) = \frac{f(t)}{S(t)} \tag{9.11}$$

Applying Equation 9.11 to first the exponential distribution we obtain as its hazard function

$$h(t) = \frac{\lambda e^{-\lambda t}}{e^{-\lambda t}} = \lambda \tag{9.12}$$

Here the hazard function is a constant; the hazard of death at any time after the time origin of the study remains the same no matter how much time has elapsed.

Next we can apply Equation 9.11 to the Weibull distribution to obtain its hazard function

$$h(t) = \frac{\lambda \gamma t^{\gamma-1} \exp(-\lambda t^{\gamma})}{\exp(-\lambda t^{\gamma})} = \lambda \gamma t^{\gamma-1} \tag{9.13}$$

By plotting this hazard function for different values of λ and γ (see Figure 9.5) we see that the Weibull distribution can accommodate increasing, decreasing and constant hazard functions. In practice, constant hazard functions are uncommon, making the exponential distribution less useful than the Weibull distribution for modelling survival times. But even the Weibull distribution may not be flexible enough for many examples of survival data as we can see from the hazard function for death in human beings given in Figure 9.6, which has a 'bath-tub' shape, being relatively high immediately after birth, declining rapidly in the early years

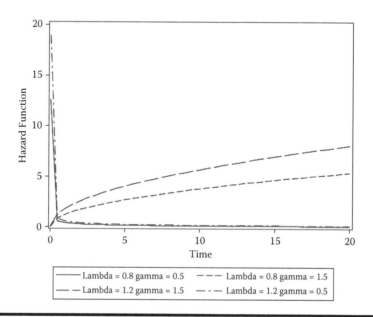

Figure 9.5   Weibull hazard functions.

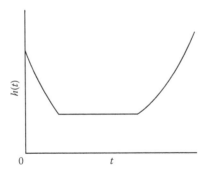

Figure 9.6 Bath-tub hazard.

and then remaining relatively constant before beginning its inexorable rise in the later years.

The hazard function can be estimated from sample data as the proportion of individuals experiencing the event of interest in an interval per unit time, given that they have survived to the beginning of the interval, i.e.

$$\hat{h}(t) = \frac{\text{number of individuals 'dying' in the interval beginning at time } t}{(\text{number of individuals alive at time } t)(\text{interval width})} \quad (9.14)$$

The sampling variation in the estimate of the hazard function within each interval is usually considerable and in practice sample hazard functions are rarely useful and consequently are rarely plotted.

But the hazard function remains of great importance for survival data because the concept is the basis of the most commonly used model for assessing the effects of covariates on survival times, namely the *proportional hazards model*, also often known as *Cox regression* after its inventor Sir David Cox. A brief account of the model follows.

9.5 Cox Regression

In considering models for survival data the first question that needs to be addressed and satisfactorily answered is: What are we going to model? More specifically, what is going to play the role of the systematic component in a regression model? According to Hosmer and Lemeshow (2000), it is the inherent aging process that is present when individuals are followed over time that distinguishes survival times from other response (dependent) variables, and it is the hazard function that most directly captures the essence of this aging process. Consequently it is natural to consider regression models for the hazard function in the analysis of survival time data. And the most common of these models is what is known as the *proportional hazards model* or *Cox regression*.

To begin we will suppose that survival data have been collected on n individuals in the context of a clinical trial and that there is only a single covariate of interest and that is treatment group: coded 0 for the standard treatment group and 1 for the new treatment group. Suppose $h_0(t)$ and $h_1(t)$ are the corresponding hazard functions for the two groups, then the proportional hazards assumption implies that

$$h_1(t) = \psi h_0(t) \tag{9.15}$$

where ψ is a constant giving the ratio of the hazards of death at any time for an individual on the new treatment relative to one on the standard treatment; ψ is known as the *relative hazard* or the *hazard ratio*. If $\psi < 1$, the hazard of death at t is smaller for an individual on the new treatment, relative to an individual on the standard treatment and if $\psi > 1$ the reverse is the case. Taking logarithms of both sides of Equation 9.15 produces the following equation

$$\log[h_1(t)] = \log(\psi) + \log[h_0(t)] \tag{9.16}$$

The proportional hazards function means that if graphs were drawn of $\log[h_1(t)]$ and $\log[h_0(t)]$, then regardless of how complex (or indeed, how simple) was the baseline hazard function the vertical distance between the two curves at any point in time will be $log(\psi)$. An implication of the proportional hazards assumption is that the population survivor functions for the two groups do not cross. Proportionality of hazards is an assumption that needs to be checked; suitable methods will be described later.

As ψ cannot be negative we can write it as $exp(\beta)$, where the parameter β is the log of the hazard ratio. Note that with the coding used for treatment group, positive values of β are obtained when ψ is greater than one, i.e. when the new treatment is inferior to the standard. By introducing an explanatory variable, x_i, for treatment group for the ith individual, with values one and zero for new and standard treatment, respectively, the hazard function for this individual, $h_i(t)$ can be written as

$$h_i(t) = e^{\beta x_i} h_0(t) \tag{9.17}$$

This model can be extended to the situation where there are q covariates measured at the start of the study, which for the ith individual take the values $\mathbf{x}_i' = [x_{i1}, x_{i2}, \ldots, x_{iq}]$ with these covariates allowed to be a mixture of continuous and binary variables (and also categorical variables with more than two categories if suitably coded as a series of dummy variables). The model is now

$$h_i(t) = e^{\left[\beta_1 x_{i1} + \beta_2 x_{i2} + \ldots + \beta_q x_{iq}\right]} h_0(t) \tag{9.18}$$

In this model the regression coefficient, $exp(\beta_j)$, gives the relative hazard for two individuals differing by one unit on the jth covariate, with all other covariates being the same for the two individuals. Now $h_0(t)$ is known as the *baseline hazard function* and is the hazard function for an individual with zero values for all covariates, or if the covariates are re-expressed as differences from their mean values, the hazard function of an individual with the mean value of each covariate (see later for more details). The model in Equation 9.18 can be written in the form

$$\log\left[\frac{h_i(t)}{h_0(t)}\right] = \beta_1 x_{i1} + \beta_2 x_{i2} + \ldots + \beta_q x_{iq} \tag{9.19}$$

So the proportional hazards function may be regarded as a linear model for the logarithm of the hazard ratio.

Before the model can be of any use in the analysis of survival data, we will, of course, need to estimate its parameters, $\beta' = [\beta_1, \beta_2, \ldots \beta_q]$. If we are willing to assume that the observed survival times are taken from a population with a particular distribution then we can use maximum likelihood estimation. For example, if we assume the survival times arise from a Weibull distribution for which the hazard function is $\lambda \gamma t^{\gamma-1}$ then the hazard function for the ith individual will be

$$h_i(t) = e^{\left[\beta_1 x_{i1} + \beta_2 x_{i2} + \ldots + \beta_q x_{iq}\right]} \lambda \gamma t^{\gamma-1} \tag{9.20}$$

Maximum likelihood estimation can now be applied to find estimates of the regression coefficients, β and the parameters, λ and γ; see, for example, Collett (2004). But there are two problems with this approach:

1. Small or even moderate-sized samples of survival times often give little evidence about the form that it is reasonable to assume for their distribution
2. Hazard functions met in practice are unlikely to always be of the simple increasing or decreasing types implied by the Weibull assumption. They may often be more complex, see, for example, the 'bath-tub' hazard function earlier in the chapter.

For these reasons Sir David Cox in his classic 1972 paper (Cox, 1972) developed an approach, now generally called simply *Cox regression*, in which the regression coefficients in Equation 9.19 can be estimated without making any assumptions about the form of the baseline hazard and therefore inferences about the effects of the covariates on the relative hazard can be made without the need for an estimate of $h_0(t)$. Cox regression is a *semi-parametric model*; it makes a parametric assumption concerning the effect of the predictors on the hazard function, but makes no assumption regarding the nature of the hazard function itself. In many situations, either the form of the true hazard function is unknown or it is complex and most interest centres on the effects of the covariates rather than the exact nature of the hazard function. Cox regression allows the shape of the hazard function to be ignored when making inferences about the regression coefficients in the model.

Estimation for Cox regression involves a procedure known as *partial likelihood*; the essence of this approach is that the partial likelihood function depends only on the vector of regression coefficients, β, *not* on the baseline hazard. Details are given in Kalbfleisch and Prentice (1980) and Collett (2004).

Interpretation of a fitted Cox regression and assessing whether or not assumptions such as the proportionality of the hazards are valid can often be helped by various types of graphical displays, as we shall attempt to show in the following sub-sections.

9.5.1 Cox Regression and the Breast Cancer Data

In Der and Everitt (2013) it was shown that a suitable, parsimonious Cox regression model for the breast cancer data was one that included only the variables hormone, tumour grade and number of positive nodes. This model was chosen by a backwards elimination procedure applied to the model that included all the covariates (again, see Der and Everitt, 2013, for details). The estimated regression coefficients for the selected model are shown in Table 9.3.

The results show that the hazard of death for patients having the hormonal therapy is estimated to be 0.71 times the corresponding hazard for patients not having the treatment when the other variables are kept constant with a 95% confidence interval of [0.56, 0.91].

Table 9.3   Results for the Cox Regression Model Fitted to the German Breast Cancer Data

Analysis of Maximum Likelihood Estimates					
Parameter	DF	Parameter Estimate	Standard Error	Chi-Square	Pr > ChiSq
horm	1	−0.33739	0.12554	7.2227	0.0072
tgrade I	1	−1.02709	0.26243	15.3180	<.0001
tgrade II	1	−0.25250	0.13348	3.5786	0.0585
pnodes	1	0.05531	0.00679	66.4346	<.0001

Analysis of Maximum Likelihood Estimates			
Parameter	Hazard Ratio	95% Hazard Ratio Confidence Limits	Label
horm	0.714	0.558	0.913	
tgrade I	0.358	0.214	0.599	tgrade I
tgrade II	0.777	0.598	1.009	tgrade II
pnodes	1.057	1.043	1.071	

Tumour grade III is the reference category and so the results in Table 9.3 show that, for example a tumour of grade I implies a hazard of death of between 0.21 and 0.60 of the corresponding hazard for a patient with tumour grade III when the other variables are unchanged.

Finally an increase of one in the number of positive lymph nodes produces an increase in the hazard of death of between 4% and 7%.

The baseline hazard function can be estimated as shown in Der and Everitt (2013) and can then be used to give an estimate of the baseline survivor function, $\hat{S}_0(t)$. From the latter the estimated survivor functions for the individuals in the sample can be found. For example, for the ith individual with a vector of covariate values, $\mathbf{x}_i$, the estimated survivor function is

$$\hat{S}(t) = \left[\hat{S}_0(t)\right]^{\exp(\hat{\beta}'x_i)} \tag{9.21}$$

(See Collett, 2004, for full details.)

The default baseline survivor function calculates the function at the mean values of numerical covariates and for the reference categories of categorical covariates. This can be found and plotted using **proc phreg** as follows:

```
proc phreg data=GBSG2 plots=survival;
   class tgrade;
   model mnths*status(0)=horm tgrade pnodes/rl;
   baseline;
run;
```

The **plots** option on the proc statement is used in conjunction with the **baseline** statement for this plot. On the **model** statement the survival time and censoring are specified in the same way as for the **proc lifetest**.

The result is shown in Figure 9.7 and shows the gradual decline in the estimated survivor function as time passes.

But in most cases where Cox regression is used, the baseline hazard function calculated in this way will not be the main focus of interest. Of far more interest will be the estimated survivor functions of individuals with other covariate values. To estimate the survivor function for other covariate values the **baseline** statement is used in conjunction with a covariates dataset. We first illustrate this procedure to get estimated survivor functions for patients who have and who have not been given hormonal treatment at the median number of positive lymph nodes and for tumour grade III. The required code is

```
data covs;
input tgrade $ pnodes horm;
cards;
III 3 0
III 3 1
;
```

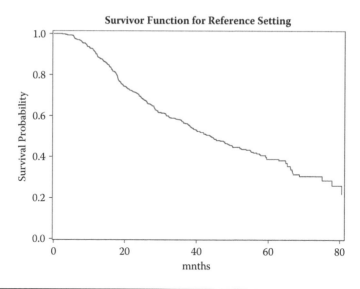

Figure 9.7 **Estimated baseline survival function at the mean values of numerical covariates and for the reference categories of categorical covariates.**

```
proc phreg data=GBSG2 plots(overlay cl)=s;
    class tgrade;
    model mnths*status(0)=horm tgrade pnodes/rl;
    baseline covariates=covs/id=horm;
run;
```

A short datastep creates a small dataset with one observation for each covariate pattern to be plotted. This dataset must contain all the same variables as the proc phreg model. This dataset is then specified with the covariates= option on the baseline statement. The covariates dataset can also contain a variable to label the covariate patterns via the id= option. Here we use the horm variable. The plots option on the proc statement is modified with plot options for confidence limits and to overlay the separate survivor functions on one plot.

The result is shown in Figure 9.8. The survivor function for the patients who have had the treatment is higher throughout the time of the study; for example the median survival time for untreated patients is approximately 40 months and for treated patients it is approximately 60 months.

A second illustration is provided by finding and plotting the estimated survivor functions of patients with different tumour grades, median number of positive lymph nodes and on hormone treatment.

```
data covs;
input tgrade $ pnodes horm;
cards;
```

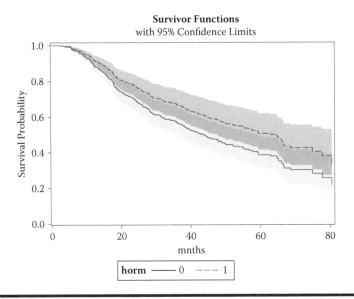

Figure 9.8 **Estimated survivor functions for treated and untreated patients at the median value of pnodes and grade III tumours.**

```
I 3 1
II 3 1
III 3 1
;

proc phreg data=GBSG2 plots(overlay cl)=s;
   class tgrade;
   model mnths*status(0)=horm tgrade pnodes / rl;
   baseline covariates=covs / id=tgrade;
run;
```

The main change has been to redefine the **covs** dataset with different values with a different value for the **id** option on the baseline statement.

The result is shown in Figure 9.9 and shows that patients with tumour grade I have much better survival experience than those with the other two grades but little difference between survival for the latter two groups.

9.5.2 *Checking Assumptions in Cox's Regression*

After any statistical model has been fitted to a dataset and the parameters of the model estimated, the adequacy of the model needs to be assessed to ensure that inferences drawn are defendable and valid. A thorough examination of how well the assumptions made by a model are met is just as important as systematic development of a model. When modelling survival data and similar to dealing

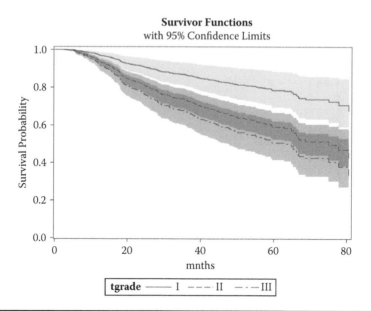

Figure 9.9 **Estimated survivor functions for patients with different tumour grades, median number of positive lymph nodes and on hormone treatment.**

with the linear regression models (see Chapter 6), assessment of model adequacy is based on the use of *residuals*. But the residuals for survival data are not nearly as obvious as those used in multiple regression and the absence of an obvious residual has led to several having been proposed which are defined explicitly in Der and Everitt (2013); here we will simply show some plots of two such residuals, *Cox–Snell* and *Deviance*, for the chosen model fitted to the breast cancer data.

9.5.2.1 Cox–Snell Residuals

If the correct model has been fitted, the Cox–Snell residuals for the n individuals in the sample will be n observations from a unit exponential distribution. (In fact, if an observed survival time is right censored, then the corresponding residual is also right censored and the residuals will be a censored sample from the exponential distribution.)

To check that the Cox–Snell residuals do have the appropriate exponential distribution, Collett (2004) shows that if the Kaplan–Meier estimate of the survivor function of the residuals is computed (denoted by $\hat{S}(r_i^{(CS)})$) with residuals from censored observations themselves regarded as censored and the values of $\log\{-\log \hat{S}(r_i^{(CS)})\}$ are plotted against the values of $\log r_i^{(CS)}$, then a straight line plot with unit slope and zero intercept will indicate that the fitted model is correct.

Systematic departures from this straight line or a straight line that does not have unit slope or zero intercept suggests that the model needs to be modified in some way. The required plot can be constructed from the following SAS code:

```
proc phreg data=GBSG2;
   class tgrade;
   model mnths*status(0)=horm tgrade pnodes;
   output out=phout logsurv=ls;
run;
data phout;
   set phout;
   rcs=ls*-1;
run;
proc phreg data=phout noprint;
   class tgrade;
   model rcs*status(0)=;
   output out=phout2 survival=srcs/method=pl;
run;
data phout2;
   set phout2;
   llsrcs=log(-1*log(srcs));
   lrcs=log(rcs);
run;

proc sgplot data=phout2;
   reg y=llsrcs x=lrcs/legendlabel='Observed line';
   refline 0;
   refline 0/axis=x;
   lineparm x=0 y=0 slope=1/legendlabel='Expected line';
   yaxis label="log-logS(r)";
   xaxis label="logr";
   where status=1;
run;
```

The first **proc phreg** step estimates and saves log survival values, then a short datastep calculates the Cox–Snell residual. These are then used as the data for the second **proc phreg** step to estimate the survival function of the Cox–Snell residuals. The second datastep calculates the values to be plotted.

The required plot at its barest would require only the **reg** plot statement within **proc sgplot**, which both produce a scatterplot and fit the regression line. Here the basic plot has been enhanced in a number of ways. Two **refline** statements add reference lines to the *x* and *y* axes at zero so that the origin is marked. The **lineparm** statement adds a line that runs through the origin with a unit slope for comparison with that fitted by the **reg** statement. The **legendlabel** option is used on both the **reg** and **lineparm** statements to give an informative legend. Finally, the *x* and *y* axes are given meaningful labels. The **where** statement restricts the plot to cases

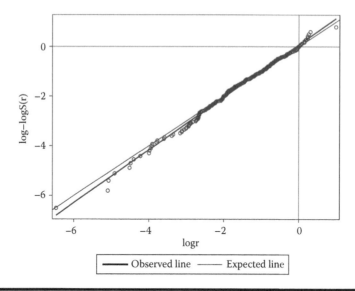

Figure 9.10 Plot of the logarithm of the Cox–Snell residuals and their log-log transformed probabilities for the breast cancer data.

with observed survival times, as the Cox–Snell residuals are not comparable for censored observations.

The resulting plot is shown in Figure 9.10. The plot shows no obvious cause for concern.

9.5.2.2 Deviance Residuals

Deviance residuals can be used to assess whether any particular individuals are poorly fitted by the chosen model where a large negative or positive value of the residual indicates a lack of fit. And deviance residuals can also be plotted against the corresponding values of a continuous covariate to investigate the appropriate functional form of the variable in the model.

We can use the **output** statement to calculate the deviance residuals and begin by plotting them against the values of each of the covariates in the model.

```
proc phreg data=GBSG2;
    class tgrade;
    model mnths*status(0)=horm tgrade pnodes/rl;
    output out=phout resdev=dres;
run;

proc sgscatter data=phout;
    plot dres*(pnodes horm tgrade)/columns=2;
run;
```

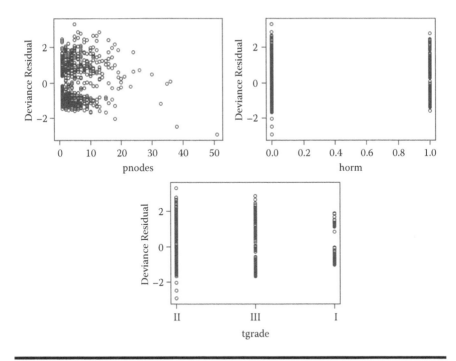

Figure 9.11 Plots of deviance residuals for the breast cancer data.

The plot is shown in Figure 9.11. The plot for the number of positive lymph nodes shows no discernible pattern, but there are perhaps two residuals that stand out from the rest, both with values less than −2. These correspond to observations with **pnode** values of 38 and 51. It may be sensible to examine how the fitted model changes if these two observations are removed, but this is left as an exercise for the reader (see Exercise 9.2).

The residuals for the two treatment groups appear to have very similar distributions, as do those for the three tumour grades. Overall the deviance residual plots indicate only the problem with the possible outliers in the data.

9.5.3 Checking the Proportional Hazards Assumption and the Appropriate Functional Form of Covariates

The validity of the results from a Cox regression depends on the proportional hazards assumption being true, i.e. that the hazard ratio is time invariant. To check the proportional hazards assumption we can use a method suggested in Lin et al. (1993), which involves the use of what are known as *martingale residuals* and which can also be used to check the appropriate functional form of the covariates, for example, whether or not including a covariate as a linear effect is acceptable. The procedures involved are complex and involve plotting a particular function of the martingale residuals against time to check the proportional hazards function and a different function against the values of the covariate of interest to investigate the assumption made about the functional form of the covariate. Observed paths in these plots are then compared to a large number of paths resulting from simulations from the model when the proportional hazards assumption holds or where the correct functional form for a covariate is being used.

Both types of plot can be invoked via the **assess** statement as we shall now illustrate by re-running the previous **proc phreg** step adding

```
assess ph var=(pnodes)/resample;
```

Using **ph** on the **assess** statement tests the proportional hazards assumption for *all* variables in the model and a separate plot is produced for each variable. Only the plot for **tgrade** I is included here and is shown in Figure 9.12. The **var=** option is used to test the functional form of one or more variables; here we test that of **pnodes**, which is included in the model as a linear effect, but a list of variables could be included in the parentheses. The **resample** option requests Kolmogorov supremum tests (see Lin et al., 1993) based, by default, on 1000 simulations and the result of the test is included as an inset in the resulting plot, which is shown in Figure 9.13.

Looking first at Figure 9.12 we see that the observed path falls well outside the envelope of simulated paths indicating that the assumption of proportional hazards for tumour grade is *not* justified. This suggests that is may be sensible to consider a

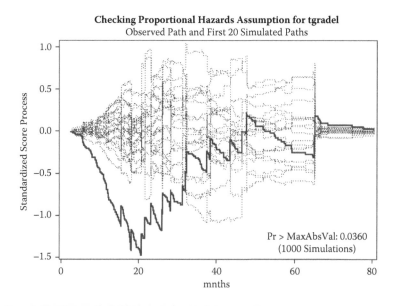

Figure 9.12 **Testing the assumption of proportional hazards for tgrade I in the breast cancer data.**

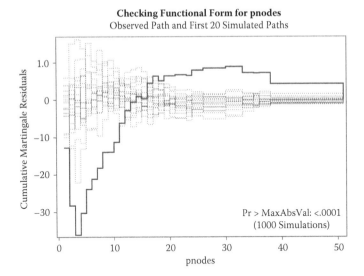

Checking Functional Form for pnodes
Observed Path and First 20 Simulated Paths

Figure 9.13   Testing the functional form of pnodes in the breast cancer data.

stratified Cox regression, with the stratifying variable being grade of tumour. Details are given in Der and Everitt (2013) and we leave it as an exercise for the reader (see Exercise 9.4).

Moving on to Figure 9.13 we again see that the observed path for **pnodes** is outside the envelope of simulated pathways and the particular pattern seen, with values below the envelope for low values of **pnodes** and values above the envelope for higher values of **pnodes** indicates, according to Lin et al. (1993), that **pnodes** should be used in the Cox regression model as **log(pnodes)**. Again we suggest the reader investigate this possibility (see Exercise 9.3). After fitting the new model the corresponding plot to that given in Figure 9.10 needs to be examined to assess its linearity or otherwise (again, see Exercise 9.3).

9.6 Graphical Presentation of Results from an Analysis of Survival Data

The results of a Cox regression analysis are often more effectively conveyed graphically, particularly for presentations to, for example, clinicians. To do this, the estimated hazard ratios and their confidence intervals from the model can be saved in a dataset, as follows:

```
proc phreg data=GBSG2;
   class tgrade;
   model mnths*status(0)=horm tgrade pnodes/rl;
   hazardratio tgrade;
```

```
    hazardratio horm;
    hazardratio pnodes/units=5;
    ods output HazardRatios=hrs;
run;
```

We include a **hazardratio** statement in the **proc phreg** step for each effect that we want to plot. For categorical variables, such as **tgrade**, the default is to produce hazard ratios comparing all of the categories; the alternative, specified via the **diff=ref** option, would be to compare all categories with the reference category. For continuous variables the default is to estimate the hazard ratio for one unit change, but this often leads to hazard ratios that appear very small, i.e. close to one. Using the **units=** option on the **hazardratio** statement allows hazard ratios to be estimated for differences of several units without having to recode the variable. Here we specify that the hazard ratio for **pnodes** is for an increase of 5 positive lymph nodes. The **ods output** statement saves all the hazard ratios and their confidence limits in the dataset **hrs**. We then use this dataset to produce a plot of the results as follows:

```
proc sgplot data=hrs noautolegend;
    highlow y=description high=waldupper low=waldlower;
    scatter y=description x=hazardratio/markerattrs=(symbol=circlefilled);
    refline 1/axis=x;
    yaxis type=discrete label='Comparison';
    xaxis Label='Hazard Ratio';
run;
```

The **hrs** dataset contains four variables: **description, hazardratio, waldupper** and **waldlower**. The **description** is of the comparison involved in the hazard ratio. The Wald confidence limits are the default. The confidence limits are plotted with a **highlow** statement (normally used for stock market price data) and the hazard ratios are overlaid using a **scatter** statement. A filled circle is used for the plotting symbol, as this looks better than a hollow circle with the line running through it. The *y* variable is not a number, as it usually is, but a character variable and so the *y* axis is treated as discrete. The option **type=discrete** is, therefore, not strictly needed. The **discreteorder=** option on the **axis** statement can be useful for controlling the order in which categorical values are plotted. A reference line is put at 1 on the *x* axis to indicate the null value.

The resulting plot is shown in Figure 9.14. This clearly demonstrates that the hormone treatment reduces the hazard of death in these patients, that having five more pnodes than another patient significantly increases the hazard of death and that grade I tumours have a lower hazard of death than tumours of both grade II and III; the hazard ratio confidence interval for tumours of grade II and grade III suggests that patients with grade II tumours will have a better survival experience than those with a grade III tumour but the confidence interval does contain the 'no difference' value of one.

One situation where the vast majority of audiences will find a graphical illustration of the results more accessible is where there is an interaction, particularly one

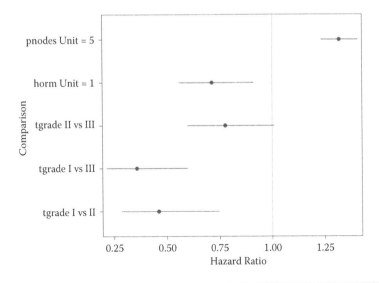

Figure 9.14   Presentation of results from the Cox regression model for the breast cancer data.

involving a continuous predictor. Further investigation into a suitable Cox regression model for the breast cancer data suggests that there is a significant interaction between hormone treatment and number of positive lymph nodes. We can fit the corresponding model and illustrate the results as follows:

```
proc phreg data=GBSG2 plots=survival;
   class tgrade;
   model mnths*status(0)=horm tgrade pnodes pnodes*horm / rl;
   hazardratio horm / at (pnodes=1 3 7 13);
   ods output HazardRatios=hrs;
run;

proc sgplot data=hrs noautolegend;
   highlow y=description high=waldupper low=waldlower;
   scatter y=description x=hazardratio / markerattrs=(symbol=circlefilled);
   refline 1 / axis=x;
   yaxis type=discrete label='Comparison';
   xaxis Label='Hazard Ratio';
run;
```

The key point here is that the **hazardratio** statement calculates the ratio for **horm** at specific values of **pnodes**. The result is shown in Figure 9.15. This plot suggests that the positive effect of the treatment on survival lessens as the number of **pnodes** increases and that the treatment becomes ineffective when **pnodes** takes the value 13.

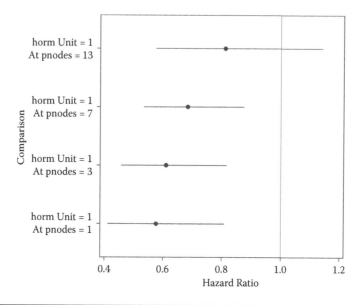

Figure 9.15 Hazard ratios for treatment effect at different values for pnodes.

9.7 Summary

Survival analysis is the study of the distribution of times to some terminating event (death, relapse etc.). A distinguishing feature of survival data is the presence of censored observations and this has led to the development of a wide range of methodology for analysing survival times. Of the available methods, Cox's regression, which allows the investigation of the effects of multiple covariates on the hazard function, is the one most commonly applied. The model has been almost universally adopted by statisticians and applied researchers primarily because it allows inferences about the regression coefficients without making any assumptions about the shape of the baseline hazard. But interpretation of the numerical results from a Cox regression is likely to be made far simpler if appropriate graphical material accompanies these results; additionally the graphics may help communicate results to the non-statisticians who have collected the data.

Exercises

9.1 The data in Table 9.4 show the results of a non-randomised clinical trial investigating a novel radioimmunotherapy (RIT) in malignant glioma patients (see Grana et al., 2002). A control group underwent the standard therapy and another group of patients was treated with RIT added.

Table 9.4   Survival Data for Patients Suffering from Two Types of Glioma Treated with Standard Therapy or a Novel Radioimmunotherapy (RIT)

ID	Age	Sex	Group	Died	Weeks
1	41	Female	RIT	TRUE	53
2	45	Female	RIT	FALSE	28
3	48	Male	RIT	FALSE	69
4	54	Male	RIT	FALSE	58
5	40	Female	RIT	FALSE	54
6	31	Male	RIT	TRUE	25
7	53	Male	RIT	FALSE	51
8	49	Male	RIT	FALSE	61
9	36	Male	RIT	FALSE	57
10	52	Male	RIT	FALSE	57
11	57	Male	RIT	FALSE	50
12	55	Female	RIT	FALSE	43
13	70	Male	RIT	TRUE	20
14	39	Female	RIT	TRUE	14
15	40	Female	RIT	FALSE	36
16	47	Female	RIT	FALSE	59
17	58	Male	RIT	TRUE	31
18	40	Female	RIT	TRUE	14
19	36	Male	RIT	TRUE	36
1	27	Male	Control	TRUE	34
2	32	Male	Control	TRUE	32
3	53	Female	Control	TRUE	9
4	46	Male	Control	TRUE	19
5	33	Female	Control	FALSE	50
6	19	Female	Control	FALSE	48

Continued

Table 9.4 (*Continued*) Survival Data for Patients Suffering from Two Types of Glioma Treated with Standard Therapy or a Novel Radioimmunotherapy (RIT)

ID	Age	Sex	Group	Died	Weeks
7	32	Female	Control	TRUE	8
8	70	Male	Control	TRUE	8
9	72	Male	Control	TRUE	11
10	46	Male	Control	TRUE	12
11	44	Male	Control	TRUE	15
12	83	Female	Control	TRUE	5
13	57	Female	Control	TRUE	8
14	71	Female	Control	TRUE	8
15	61	Male	Control	TRUE	6
16	65	Male	Control	TRUE	14
17	50	Male	Control	TRUE	13
18	42	Female	Control	TRUE	25

Plot the estimated survivor functions for the treated and control groups separately for the two grades of tumour. What conclusions do you draw about the two treatments?

9.2 Rerun the Cox regression for the breast cancer data leaving out the two observations identified in the text as possible outliers and compare the results with those given in Table 9.3.

9.3 Returning to the non-stratified Cox model for the breast cancer data, fit the model using log(pnodes) and investigate whether this functional form for this covariate is more appropriate than simply pnodes. Construct the corresponding plot to that in Figure 9.10 for the new model. What do you conclude? When log(pnodes) is used as an explanatory variable does the model require a Treatment × Log(pnodes) interaction?

9.4 Fit a stratified Cox model to the breast cancer data using tumour grade as the stratifying variable and compare your results with those given in the text.

References

Agresti, A. (1996), *An Introduction to Categorical Data Analysis*, New York, Wiley.

Altman, D. G. (1991), *Practical Statistics for Medical Research*, London, Chapman & Hall.

Barnicot, N. A. and Brothwell, D. R. (1959), The evaluation of metrical data in the comparison of ancient and modern bones. In *Medical Biology and Etruscan Origins*, CIBA Foundation Symposium, G. E. W. Wolstenholme and C. M. O'Connor, eds., New York, Wiley.

Beck, A. T., Steer, A., and Brown, G. K. (1996), *Beck Depression Inventory Manual*, San Antonio, TX, The Psychological Corporation.

Bennett, G. W. (1988), Determination of anaerobic threshold, *Canadian Journal of Statistics*, 16, 307–310.

Bertin, J. (1981), *Semiology of Graphics*, Madison, WI, University of Wisconsin Press.

Burns, K. C. (1984), Motion sickness incidence: Distribution of time to first emesis and comparison of some complex motion conditions, *Aviation Space and Environmental Medicine*, 56, 521–527.

Cameron, E. and Pauling, L. (1978), Supplemental ascorbate in the supportive treatment of cancer: Re-evaluation of prolongation of survival times in terminal human cancer, *Proceedings of the National Academy of Science USA*, 75, 4538–4542.

Carpenter, J., Pocock, S. J. and Lamm, C. J. (2002), Coping with missing data in clinical trials: A model-based approach to asthma trials, *Statistics in Medicine*, 21, 1043–1066.

Cleveland, W. S. (1979), Robust locally weighted regression and smoothing scatterplots, *Journal of the American Statistical Society*, 74, 829–836.

Cleveland, W. S. (1993), *Visualizing Data*, Summit, NJ, Hobart Press.

Cleveland, W. S. (1994), *The Elements of Graphing Data* (Revised Edition), Summit, NJ, Hobart Press.

Cleveland, W. S. and McGill, M. E. (1987), *Dynamic Graphics for Statistics*, Belmont, CA, Wadsworth.

Collett, D. (2003), *Modelling Binary Data*, London, Chapman & Hall/CRC Press.

Collett, D. (2004), *Modelling Survival Data in Medical Research*, 2nd Edition, London, Chapman & Hall/CRC Press.

Collett, D. and Jemain, A. A. (1985), Residuals, outliers and influential observations in regression analysis, *Sains Malaysiana*, 4, 493–511.

Cook, D. and Swayne, D. F. (2007), *Interactive and Dynamic Graphics for Data Analysis*, New York, Springer.

Cook, D. and Weisberg, S. (1982), *Residuals and Influence in Regression*, London, Chapman & Hall/CRC.

225

Cox, D. R. (1972), Regression models and life tables, *Journal of the Royal Statistical Society*, Series B, 34, 187–200.

Cumming, G. (2009), Inference by eye: Reading the overlap of independent confidence intervals, *Statistics in Medicine*, 28, 205–220.

Dalal, S. R., Fowlkes, E. B. and Hoadley, B. (1989), Risk analysis of the space shuttle: Pre-Challenger prediction of failure, *Journal of the American Statistical Association*, 84, 945–957.

Der, G. and Everitt, B. S. (2009), *A Handbook of Statistical Analyses Using SAS*, Boca Raton, FL, Taylor & Francis.

Der, G. and Everitt, B. S. (2013), *Applied Medical Statistics Using SAS*, 3rd Edition, Boca Raton, FL, Taylor & Francis.

Digby, P. G. N. and Kempton, R. A. (1987), *Multivariate Analysis of Ecological Communities*, London, Chapman & Hall.

Diggle, P. L., Liang, K. and Zeger, S. L. (2002), *Analysis of Longitudinal Data*, 2nd Edition, Oxford, Oxford University Press.

Dizney, H. and Gromen, L. (1967), Predictive validity and differential achievement in three MLA comparative foreign language tests, *Educational and Psychological Measurement*, 27, 1127–1130.

Everitt, B. S. (2010), *Multivariable Modeling and Multivariate Analysis for the Behavioral Sciences*, Boca Raton, FL, CRC Press.

Everitt, B. S. and Dunn, G. (2001), *Applied Multivariate Data Analysis*, 2nd Edition, Chichester, Wiley.

Everitt, B. S. and Hand, D. J. (1981), *Finite Mixture Distributions*, London, Chapman & Hall.

Everitt, B. S., Landau, S., Leese, M. and Stahl, D. (2011), *Cluster Analysis*, 5th Edition, Chichester, Wiley.

Everitt, B. S. and Skrondal, A. (2010), *Cambridge Dictionary of Statistics*, 4th Edition, Cambridge, Cambridge University Press.

Fitzmaurice, G. M., Laird, N. M. and Ware, J. H. (2004), *Applied Longitudinal Analysis*, New York, Wiley.

Friendly, M. and Denis, D. J. (2006), *Graphical Milestones*, http://www.mayh.yorku.ca/SCS/Gallery/milestone

Galton, F. (1886), Regression towards mediocrity in hereditary stature, *Journal of the Anthropological Institute of Great Britain and Ireland*, 15, 246–263.

Goldberg, D. (1972), *The Detection of Psychiatric Illness by Questionnaire*, Oxford, Oxford University Press.

Grana, C., Chinol, M., etc (2002), REF IN SAS 3

Greenacre, M. (1992), Correspondence analysis in medical research, *Statistical Methods in Medical Research*, 1, 97–117.

Gregoire, A. J. P., Kumar, R., Everitt, B. S., Henderson, A. F. and Studd, J. W. W. (1996), Transdermal oestrogen for the treatment of severe post-natal depression, *The Lancet*, 347, 930–934.

Haberman, S. J. (1972), Log-linear fit for contingency tables—Algorithm AS51. *Applied Statistics*, 21, 218–225.

Haberman, S. J. (1973), The analysis of residuals in cross-classified tables, *Biometrics*, 29, 205–220.

Hancock, B. W., Bruce, L., May, K. and Richmond, J. (1979), Ferritin, a sensitizing substance in the leucocyte migration inhibition test in patients with malignant lymphoma, *British Journal of Hematology*, 43, 273–233.

Hilts, V. L. (1975), *A Guide to Francis Galton's English Men of Science*, Philadelphia, American Philosophical Society.

Hosmer, D. W. and Lemeshow, S. (2000), *Applied Logistic Regression*, 2nd Edition, New York, Wiley.

Hosmer, D. W., Lemeshow, S. and Sturdivant, R. X. (2013), *Applied Logistic Regression*, 3rd Edition, New York, Wiley.

Howell, D. C. (2002), *Statistical Methods in Psychology*, 5th Edition, Pacific Grove, CA, Duxbury.

Kalbfleisch, J. D. and Prentice, J. L. (1980), *The Statistical Analysis of Failure Time Data*, New York, Wiley.

Lin, D. Y., Wei, L. J. and Ying, Z. (1993), Checking the Cox models with cumulative sums of Martingale-based residuals, *Biometrika*, 80, 557–572.

MacGregor, G. A., Markandu, N. D., Roulston, J. E. and Jones, J. C. (1979), Essential hypertension: Effect of an oral inhibitor of angiotensin-converting enzyme, *British Medical Journal*, 2, 1106–1109.

Neyzi, O., Alp, H. and Orhon, A. (1975), Breast development of 318 12-13 year old Turkish girls by socio-economic class of parents, *Annals of Human Biology*, 2(1), 49–59.

Pearson, K. and Lee, A. (1903), On the laws of inheritance in man: I. Inheritance of physical characters, *Biometrika*, 2, 357–462.

Proudfoot, J., Glodberg, D., Mann, A., Everitt, B. S., Marks, I. and Gray, J. (2003), Computerised, interactive, multimedia cognitive behavioural therapy for anxiety and depression in general practice, *Psychological Medicine*, 33, 217–227.

Rawlings, J. O., Sastry, G. P. and Dickey, D. A. (2001), *Applied Regression Analysis: A Research Tool*, New York, Springer.

Sauerbrei, W. and Royston, P. (1999), Building multivariable prognostic and diagnostic models: Transformation of the predictors by using fractional polynomials, *Journal of the Royal Statistical Society, Series A*, 162, 71–94.

Schmid, C. F. (1954), *Handbook of Graphic Presentation*, New York, Ronald Press.

Schumacher, M., Basert, G., Bojar, H., Hubner, K., Olschewski, M., Sauerbrei, W., Schmoor, C., Meumann, R. L. A. and Rauschecker, H. F., for the German Breast Cancer Group (1994), Randomized 2 x 2 trial evaluating hormonal treatment and the duration of chemotherapy in node-positive breast cancer-patients, *Journal of Clinical Oncology*, 12, 2086–2093.

Silverman, B. (1986), *Density Estimation in Statistics and Data Analysis*, London, Chapman & Hall/CRC.

Sokal, R. R. and Rohlf, F. J. (1981), *Biometry*, 2nd Edition, San Francisco, W.H. Freeman.

Spilich, G. J., June, L. and Renner, J. (1992), Cigarette smoking and cognitive performance, *British Journal of Addiction*, 87, 1313–1326.

Tufte, E. R. (1983), *The Visual Display of Quantitative Information*, Cheshire, CT, Graphics Press.

Unwin, A., Theus, M., and Hofmann, H. (2006), *Graphs for Large Datasets: Visualizing a Million*, New York, Springer.

Vanisma, F. and De Greve, J. P. (1972), Close binary systems before and after mass transfer, *Astrophysics and Space Science*, 87, 377–401.

Vetter, B. M. (1980), Working women scientists and engineers, *Science*, 207, 28–34.

Vuilleumier, F. (1970), Insular biogeography in continental regions. I. The Northern Andes of South America, *American Naturalist*, 104, 373–388.

Wainer, H. (1997), *Visual Revelations*, New York, Springer.

Wood, S. N. (2006), *Generalized Additive Models: An Introduction with R*, Boca Raton, FL, Chapman & Hall/CRC.

Woodley, W. L., Simpson, J., Biondrini, R. C. and Berkeley, J. (1977), Rainfall results 1970–1975: Florida area cumulus experiment, *Science*, 195, 735–742.

Zerbe, G. O. (1979), Randomization analysis of the completely randomized design extended to growth and response curves, *Journal of the American Statistical Association*, 74, 215–221.

Zerbe, G. O. and Murphy, J. R. (1986), On multiple comparisons in the randomization analysis of growth and response curves, *Biometrics*, 42, 795–804.

Index